SOWING IDEAS.
HARVESTING THE FUTURE

Delius Klasing Verlag

CLAAS
VARIO

CLAAS
8900
LEXION
LEXION
30
30

HORSCH
HORSCH

XERION
CLAAS
5000
XERION
CLAAS

CLAAS
CLAAS
CARGOS

Why we need innovations

If we are to feed a continuously growing human population sustainably in future, we must take a joined-up approach to economics, ecology and society, says Carl-Albrecht Bartmer, long-standing former president of the German Agricultural Society (DLG). To maintain a balance between the three elements of this triangle in the face of increasing challenges, agricultural engineering must become more innovative than ever before.

PHOTOS Thorsten Doerk

Farmers have a heavy responsibility for feeding the world. They will need a wealth of innovative strength to tackle the challenges that lie ahead.

"For an innovative company, having innovation in your DNA means that you never stop innovating".

CARL-ALBRECHT BARTMER

›FEEDING THE WORLD
›FORWARD-LOOKING INDUSTRY
›AGRICULTURAL ENGINEERING

The following comparison illustrates the groundbreaking impact of agricultural innovations on global society: in around 1800, before the onset of industrialisation, 75% of the working population worked on the land. Today it is only around 2%. Do these figures imply that agriculture has become less important? No, it's the sector that supplies us with food. And it will continue to do so in the foreseeable future. But what these figures do show is that to do this job, agriculture today needs only a fraction of the labour force compared with around 200 years ago. This is due to better and better agricultural machinery, which has made farming more and more efficient, initially by mechanical means, then electro-hydraulically and now digitally as well. It is these innovations that allowed industry to take off in the first place, allowed towns and cities to develop into dynamic, creative centres, places where people find opportunities for education and work. The megatrend of urbanisation would not have been possible without innovations in agricultural engineering.

Sustainable systems are what's needed

If you read at an average speed, it will have taken you a good 20 seconds to read this far. In this short time, the global population will have grown by more than 50 people. This rate equates to an extra 80 million people per year – that's the same as the population of Germany. Experts predict that by 2055 we will have hit the ten billion mark. And this has consequences. Already, almost everywhere we look we can see how depleted the resources on this planet have become and how severe the human impact on earth has been. So it's only logical to conclude that the world needs sustainable systems. Humankind has no future without sustainability.

There are three dimensions to sustainability: firstly, systems must be economically sustainable because, taking agriculture as an example, this is the only way that we can continue to feed the world and guarantee the greatest possible prosperity. Secondly, they must be ecologically sustainable since otherwise we destroy the very foundations on which all life depends. Thirdly, these systems must be socially sustainable since that's the only way we can avoid social conflicts which threaten all other goals.

Economical, ecological and social – imagine these elements as the points of a triangle. If we strengthen one individual point, we upset the balance of the triangle. To apply this to the agricultural sector in concrete terms: if a farmer thinks strictly along economic lines, he jeopardises the ecological and social components. On the other hand, if he focuses solely on ecology, he risks the profitability of his farm. This analogy illustrates how difficult it is to do justice to sustainability as an overall concept. Yet there is a force which is able to elevate the entire triangle to a higher level – and that is the potential that comes from innovations.

Innovative agricultural engineering, a leap forward for all aspects of sustainability

By using new construction methods, innovative agricultural engineering makes it possible to perform tasks with greater efficiency, precision and autonomy. Protecting soil from compaction by completing tasks in good time is just as important as minimising emissions (e.g. greenhouse gases) and external inputs (e.g. for fertiliser and crop protection applications). At present we are making quantum leaps in reducing losses and safeguarding the quality of harvested crops. So what does innovative agricultural engineering mean for farmers? It ranges from autonomous control systems, sensor-based precision and ergonomic workplace design right up to completely autonomous, self-optimising processes, all of which are designed to make the job easier. Innovative agricultural engineering underpins the economic foundations of the farm: using technology to reduce the production costs per unit of harvested crop is the key feature that maintains a farm's competitive edge and thus its future viability in open global markets. To sum up, innovative machines don't simply relieve the strain on the driver, they also work more accurately and faster than a human ever could, thereby protecting the environment and boosting farm profitability.

Innovations are the only option for sustainable production systems. But they also call for a social setting that is open to innovation. Apart from modern varieties, efficient crop protection and intelligent use of fertilisers, agricultural tools that are fundamentally different from those that have gone before are essential. But although we have been quick to embrace innovations in communication and mobility, the agricultural sector tends to feel a certain nostalgia for the "good old days". And yet this was by no means more sustainable and furthermore, fed only a fraction of the people. In the agricultural industry we also need to have the social courage to change, because this has proved to be the hallmark of high performing societies over the centuries. Why should that not apply to the agricultural sector too? Anyway, we owe it to future generations to see innovations through.

Carl-Albrecht Bartmer was president of the German Agricultural Society (DLG) from 2006 to 2018. He lives with his family in Löbnitz an der Bode in Saxony-Anhalt, where he runs his grandfather's farm.

Being innovative is a never-ending job

The drivers of change include policymakers and civil society, but the service providers include in particular businesses – ones like CLAAS, that have been committed to progress from the very start. It's not easy being a slave to progress. Every day, employees in the company are involved in what economist Joseph Schumpeter termed "creative destruction": they question old things in order to be able to generate new ones. For innovative companies, having innovation in your DNA also means that you can never stop being innovative: on the very day that we unveiled the new LEXION, engineers were already working on its more efficient, more connected successor.

This relentless urge to innovate is important. For CLAAS it's all about corporate success and the future viability of a traditional family business. For global society, however it's about far more than that. Let's look at the figures again: we will only be able to feed the soon-to-be ten billion people on our planet if in agricultural engineering we intelligently combine efficiency with a sense of proportion. In other words, if we take a joined-up approach to the economic, ecological and social dimension. And in this context, forward-thinking entrepreneurs from August Claas in his time to the present generation are the true "labourers in the vineyard".

Where the future is well within reach

CLAAS E-Systems is housed on a five hectare site in Dissen down in the south of Lower Saxony. Since the company's electronics development centre was established in autumn 2017, it has been driving innovation at the interface between mechanical engineering and IT.

TEXT André Boß **PHOTOS** Lukas Kawa and Andreas Fechner

Dr Carsten Hoff, managing director of CLAAS E-Systems: "Here in Dissen, innovation is the air we breathe."

› DEVELOPMENT CENTRE
› AUTOMATION
› ASSISTANCE SYSTEMS

There's a whiff of Silicon Valley to the home of CLAAS E-Systems: the glazed building appears to be transparent from the outside. On the inside, however, it conceals a wealth of company secrets, which is not surprising when you think who works here: around 200 experts in software development and mechanical engineering work tirelessly to further increase the efficiency of agricultural machines by creating new concepts at the interface between electronics and digitalisation. They develop control units and electronic components, terminals and camera systems, automatic satellite-based steering systems and a whole host of other technical and electronic wizardry. These innovations are pooled under the abbreviation EASY, which stands for Efficient Agriculture Systems and describes what the people here in Dissen are feverishly working on: digital systems which drive machines to achieve better and better results.

Centre of excellence: Dissen brings together the development of control units, terminals, camera systems, automatic steering systems and a whole host of other solutions for an increasingly digitally connected agricultural industry under one roof.

It's early evening and the staff are gradually starting to trickle out of the building. For managing director Dr Carsten Hoff, the quieter part of his working day is about to start. In the mornings and afternoons he is involved in numerous small discussions and large meetings, including telephone and video conferences with CLAAS colleagues around the world. Hoff himself is often on the road, visiting the production plants in Le Mans or Bad Saulgau, meeting colleagues in Harsewinkel or the Berlin developers from the CLAAS subsidiary 365FarmNet, whose app functions also incorporate innovations from CLAAS E-Systems. Carsten Hoff explains why there is such a need for internal communications: "Our work covers virtually all aspects of machine development and on top of that, we are in constant contact with Sales, Marketing and Design."

Always in search of something new

With a PhD in electrical engineering, Carsten Hoff joined CLAAS in summer 2014, having previously worked for an automotive supplier. On switching to agricultural engineering, he was appointed managing director of the company's subsidiary CLAAS E-Systems. The newly constructed building in Dissen shows just how important the new Development Centre is to the company. Located on the edge of the Teutoburg Forest close to the A33 autobahn, the building is surrounded by open space. This is not intended to imply a desire to avoid neighbours: the company simply needs space to test drive the innovations under development. The site in Dissen is also a good match in logistical terms – the company headquarters in Harsewinkel lies only 20 kilometres to the north.

So what happens in Dissen behind the glass facade? "We are a development centre. This means that we are constantly exploring the latest digital technologies to see what potential for innovation they offer", explains the managing director. The centre has been nicknamed Dissen Valley in a nod to Silicon Valley in San Francisco, which has given rise to many key digital innovations over the years. But Carsten Hoff believes that alongside all the digital expertise, it's just as important to emphasise his team's skills in traditional, customer-centred agricultural engineering. "Many of our staff have close connections with farming." Some of them grew up on a farm and often return to help out at harvest time. "This expertise gives us a valuable advantage because our people can very accurately gauge the needs of our customers. Or to put it another way: no one here would allow themselves to be blinded by digital options. We never lose sight of the idea that a product has to be usable in the field."

The benefits of innovations must be immediately obvious

The product managers have a particularly important role to play here. Often agricultural graduates, it is their job to carefully investigate whether the ideas being developed at CLAAS E-Systems actually generate added commercial value. "That's essential, because CLAAS customers rightly expect our innovative features to either increase their yields or make their job easier. The knowledge and experience of our product managers prevents us from developing technology-driven innovations purely for their own sake, the benefits of which are not immediately obvious to the customer." For if there's one thing that Carsten Hoff is sure of, it's that this determines whether an innovation will be successful or not. "If we have to explain at length their added value, we dampen our customers' enthusiasm. And I believe that this emotional connection plays an important role in the purchasing decision."

Carsten Hoff picks up a tablet and presents some of the developments that his team have recently been working on. They are focusing on three areas of modern agriculture where digital solutions can really help: Connectivity & Data Management provides telematic applications which allow customers to access machine data at any time via the Internet. "This makes the machine performance transparent at all times", says Hoff. In Process Automation CLAAS E-Systems is developing applications to automate an increasing number of processes. Hoff: "This increases efficiency and relieves the burden on the driver." The work of the Steering, Precision Applications & Terminals team ranges from automatic guidance to fully automated steering systems and control units which allow farmers to control highly complex machine processes easily and intuitively. "When it comes to ease of operation, today's mobile phones have set the bar very high," says Carsten Hoff as he reaches for his smartphone: "Nowadays people expect apps, even ones which perform complex tasks, to be easy to use. And we have to keep pace with this trend."

Fast-paced digital change

Let's return to the core issue with these developments: do these innovations ultimately tip the balance when it comes to the customer choosing whether or not to buy CLAAS? Or are they merely gadgets which the customer finds useful but which hold less sway than traditional indicators such as hp output, price or fuel consumption when it comes to making a purchasing decision? Carsten Hoff grins; he knows that this question lies at the heart of modern agricultural engineering. "In reality, we recognise that the electronic features now have a decisive influence on the purchasing decision." So in the same way that a new car comes with a navigation system nowadays, a GPS steering system has become standard on machines, especially for large customers. Increasingly, it is also true of features which automatically adjust the machine settings. Especially when absolute precision and every minute counts: "In the short harvesting window, everything has to run smoothly and a machine that can take control of the most important settings avoids errors."

To achieve this, the company has to experiment, as it becomes clear when we walk through the development centre: work here is done not just in offices, but also in rooms that look more like laboratories. "Innovations are invariably associated with genuine unique selling points. So our goal is as follows: we want to develop something that brings customers

The home of CLAAS E-Systems in Dissen connects the present with the future inside and outside the building.

When innovation is on the agenda, regular meetings are essential to find out what the new solution can do and how it helps the customer.

one step further forward. We want to formulate exceptional ideas – with the aim of then incorporating them into the machine." Since the cycles for innovation in the digital sector are becoming increasingly shorter, standing still is not an option for CLAAS E-Systems. Carsten Hoff picks up his smartphone again and says: "Take my brand-new mobile phone. In two years it will be virtually an antique." Agriculture is also ruled by this pace of change: "We want to contribute to progress every year. Our customers expect this of us – and so we also expect it of ourselves."

Powerful machines are still the centrepiece

In conversation with Carsten Hoff, you get an impression of what agricultural machinery will look like in future. Digital for sure – but that's only part of it. A powerful machine will always be the centrepiece while the digital technology is there to ensure that the machine performs to its best as efficiently as possible. "Every product we develop here has a direct interface with the machine", explains Carsten Hoff. The assistance systems ensure that the machine is perfectly set up, the telematics systems send data from the machine to the web interface, the sensors installed on the front of the tractor, for example, measure the crop density and nutrient requirements. "The future will belong to machines equipped with sensors, data and built-in artificial intelligence which enables them to make better and better decisions – about when I should harvest or how I should fertilise my field, for instance." Customers then have the choice: do they let the system suggest an approach, which they then decide to accept or reject? Or do they leave it entirely up to the system to decide on an approach to apply to the whole field or predefined sections of it? "We see it as our task to support the customer in making this choice", says Hoff. "Ultimately, we are not just providing new technology but together with our partners, thinking hard about how our innovations change the way our customers' work."

A big change on the horizon is the further development of fully automated systems as we move closer to completely autonomous machines. "I expect that we will see these on the field long before they appear on the road", says Carsten Hoff. "The advantage that we have compared with the automotive industry is that our vehicles operate on private land. So many of the complex issues and regulations that car manufacturers have to grapple with don't apply." This doesn't mean that autonomous machines would have to be less than perfect. On the contrary: It's the precision of the digitally controlled steering systems that will determine whether these machines succeed, says Hoff. "Already our customers can preplan the track that the machine is to take on their computer or tablet. In this situation the driver acts as the controller. So it is also conceivable that in the near future a driver will be able to oversee diverse autonomous machines simultaneously." This would create an unprecedented degree of efficiency in farming – combined with comfort and precision. No wonder that customers enjoy hearing about these and other ideas from Dissen. At this CLAAS site the future of agricultural engineering is conceivable, tangible and even palpable. And it makes you hungry for more.

"The future will belong to machines equipped with sensors, data and built-in artificial intelligence which enables them to make better and better decisions about when I should harvest or how I should fertilise my field".

DR CARSTEN HOFF

Managing director Dr Carsten Hoff is fascinated by the opportunities presented by digital tools to make agricultural machinery even more efficient.

A revolution that comes from the heart

The task of the combine harvester has remained unchanged over the years: its job is to separate the wheat from the chaff, without losses and as quickly as possible. In over 80 years of combine harvester development, CLAAS machines have evolved from harvest helpers into giants of the fields. The new LEXION represents a real revolution in combine harvester technology: the machine is a perfectly coordinated system which brings together power, comfort and efficiency.

TEXT André Bosse **PHOTOS** CLAAS

CLAAS
8900
LEXION

The LEXION 8900 with a VARIO 1380 cutterbar harvesting rapeseed. The GPS PILOT system ensures precise guidance.

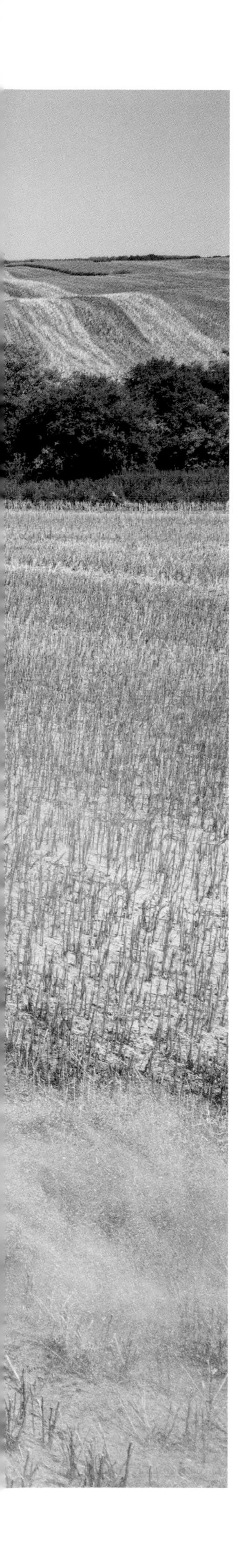

› CAPACITY FOR INNOVATION
› PRECISION
› ASSISTANCE SYSTEMS

Some 25 years ago, CLAAS launched the first combine harvester to bear the LEXION name. Today – following many more stages of development – the LEXION model series has long since earned a special place in the hearts of the development team. Time and again they improved the machine, bringing it closer to the optimum. Now CLAAS is presenting a new LEXION generation, and once again CLAAS has revolutionised the combine harvester.

The development team focused in particular on the beating heart of the machine: the new LEXION has a completely redesigned threshing unit. Since it entered series production in 1996, the LEXION has operated with APS threshing technology. The developers have now redesigned this completely for the new generation – whether in the form of the APS SYNFLOW HYBRID for the new LEXION 8000 / 7000 models or the APS SYNFLOW WALKER for the LEXION 6000 / 5000 range. The threshing drum now has a diameter of 755 mm, a whole 25% increase which optimises the crop flow. The feeder drum in the APS SYNFLOW HYBRID threshing unit is also larger than ever at 600 mm, an increase of 57%. An important innovation in the models with the APS SYNFLOW WALKER system is the large, additional separator drum with a diameter of 600 mm which enables a further increase in throughput as a result of active separation.

With the heart of the new LEXION beating stronger than ever, the machine's overall balance of performance, quality and efficiency has been maintained. The developers have made further improvements to enhance operator comfort: functions such as the hydraulic threshing concave flap or the pivoting concave bar that can be actuated from inside the cab.

Combination of power and comfort

The new LEXION combines this comfort with power and an imposing presence: the working width is up to 13.8 m, the output up to 581 kW / 790 hp, the grain tank has a volume of up to 18,000 l. For a clear view of the unloading process, the grain tank unloading tube has a pivot angle of 105°. The CLAAS developers made a point of ensuring that the power is there when the operator needs it. DYNAMIC POWER is the intelligent system which automatically adjusts the engine output to the field conditions when the engine is operating under partial load and ensures maximum efficiency at full load. The DYNAMIC COOLING system is also based on the principle of flexible performance: it provides cooling on demand, requires significantly less engine output and so helps to save fuel.

One factor which is becoming increasingly important in contemporary agriculture is precision. With harvest windows growing ever smaller, efficiency can also be achieved through accuracy which is ensured by new electronic applications. The CEMOS AUTOMATIC intelligent operator assistance system is capable of optimising the machine continuously and autonomously. The system learns for itself by trying out different settings and drawing conclusions from the results. "The automatic set-up of the machine systems of the new LEXION continues to be based on expert knowledge and the autonomous adaptation of characteristic maps to local conditions", explains Head of Development Dr Thomas Barrelmeyer. With REMOTE SERVICE, the service partners have access to the machine – but only once it has been enabled by the customer, of course. Servicing and maintenance become significantly faster and more straightforward as a result. The FIELD SCANNER, based on the latest sensor technology, eases the operator's workload: the LEXION finds the ideal track autonomously while making use of the full width of the cutterbar, thereby keeping the number of passes to a minimum. A glance at the inside of the cab confirms that the LEXION offers a wealth of functions – yet operating the machine has never been easier or more comfortable.

"We set out to look at every single aspect, every last nut and bolt, to see what could be done better".

THOMAS BARRELMEYER

54,000 optimised components

The latest LEXION generation has been engineered to attain the optimum standard in every respect without compromising the balance between performance, quality and efficiency. "The new LEXION embodies not only many new concepts, but also a large number of enhanced details", says Barrelmeyer. "We set out to look at every single aspect, every last nut and bolt, to see what could be done better." The numbers involved are huge: there are some 54,000 components in these machines. On top of those, there are the digital components based on the innovative work by the software and app developers. Today's combine harvesters are often described as "mobile factories". "But this description doesn't convey the attention the development team paid to fine-tuning the various assemblies that make up the machines", explains Barrelmeyer. "We are talking about a highly efficient and coordinated system which is designed to produce the best results and give the operator the feeling that their combine is delivering peak performance, but it's also making their task as stress-free as possible." When you stand in front of the new LEXION, climb into the cab, start it up and get to work, you can't help thinking of something a customer from the USA once said about this CLAAS combine harvester: "There's no such thing as the perfect machine, but the LEXION is as close as it gets." This feat of design engineering has been made possible on the one hand by the excellent work done by the current development team and on the other hand by the company's ability to draw on a vast wealth of experience in building combine harvesters.

A look back to the year 1936 and the world premiere of the first combine harvester from CLAAS: "Mower-Thresher-Binder" [Mäh-Dresch-Binder, MDB for short] was the name which Head Design Engineer Walter Brenner's team of developers gave their machine, a combination of a self-binder and a threshing unit. After cutting, the machine fed the straw to the threshing drum from where a chain conveyor re-routed the straw and fed it to the straw walker. After two cleaning phases, the resulting grain was filled into sacks and the straw was tied. Job done!

If this historic MDB were to be placed in a field next to the new LEXION models, the differences could hardly be more striking: here, an historic, pioneering achievement which demonstrated more than 80 years ago that the European market was ready for the introduction of combine harvesters. There, a gigantic machine embodying state-of-the-art agricultural technology, powerful and efficient, user-friendly and equipped with digital tools. The two machines are as different from each other as a dial telephone and the latest smartphone. And yet both are essentially designed to deal with the same question that has always presented farmers with a challenge: how to get the grains out of the ears as efficiently as possible.

Customer-driven innovation

Then, as now, the CLAAS developers were subject to two forces which drove them in their continuous and highly dedicated pursuit of new solutions. First of all, there are the customers, who have always had a very close relationship with their combine harvesters. As a rule, farmers and contractors do not regard their CLAAS combine harvester as just another piece of farm machinery: its sheer size, its technical equipment and the importance of its task mean that the operators develop a strong emotional attachment to their machines. And if something is important to you, you take an interest in its development. "Our in-depth dialogue with our customers tells us which innovations they would like to see and what their requirements are", says Barrelmeyer. Without a doubt, the commitment of the customers has contributed to the evolution of CLAAS combine harvesters over the past 80-plus years.

A second important innovation driver is the competition. This is demonstrated by taking a look at the period after the Second World War: at that time, when the CLAAS SUPER model

"Proving how good we are".

The LEXION 8900 with a VARIO 1230 cutterbar harvesting wheat.
The adjustment options of the VARIO cutterbar ensure that even laid crops are harvested cleanly.

some **54,000 components**
13 new models with 313 to 790 hp
up to **13,8 m working width**
up to **25% greater harvesting performance**
1 l of diesel to harvest 1,000 kg of grain
up to **18,000 l capacity,** equivalent to 128 baths full
grain tank unloading in 100 seconds, equivalent to one bath full per second
up to **40 km/h** ground speed

Klaus Schäfer, Product Manager CLAAS Self-propelled Harvesters

For LEXION Product Manager Klaus Schäfer, the company's history of innovation plays a decisive role in gaining customer acceptance for developments such as the new LEXION.

Mr Schäfer, to what extent has the company's capacity for innovation influenced the development of the new LEXION combine harvester?

The new LEXION straw walker machines with the APS SYNFLOW WALKER threshing unit have achieved a performance increase of 25%. That means that the farmer can use the machine to harvest a quarter more in the same time. What's more, compared with the first LEXION of 1996, the machine has a raft of additional automation and digitalisation features. These make for lower fuel consumption and improved grain quality, to mention just two examples.

How do you convince customers of the benefits of your innovations?

The customer has to try out our products and then want for nothing. Our tradition of innovation is equally important: we have to provide the evidence which forms the basis for the customer to have confidence in us. For example, CLAAS has always been the leader with regard to combine harvester throughput. We are also the ones who established technologies such as TERRA TRAC or our CEMOS AUTOMATIC operator assistance system in the market. And we were the first to offer a membrane keyboard in the cab or touchscreens – which are unaffected by a tough working environment – with CEBIS TOUCH. This historical and technical evidence of our capacity for innovation forms the basis for us to convince younger farmers, too.

The new LEXION can harvest huge areas in a very short time. Is that what the future of agriculture is all about?

Much of it will continue to follow this pattern for a long time. But we are also looking at completely different topics in order to respond to the wishes of our customers – and of consumers, too. Sustainability and traceability, as well as new cultivation methods, are becoming increasingly relevant. We position our machines to meet customers' wishes – and we demonstrate this anew, year after year.

The cab of the LEXION offers excellent driving comfort and a control concept that meets every operator's requirements.

All engine variants are equipped with DYNAMIC POWER for active engine control.

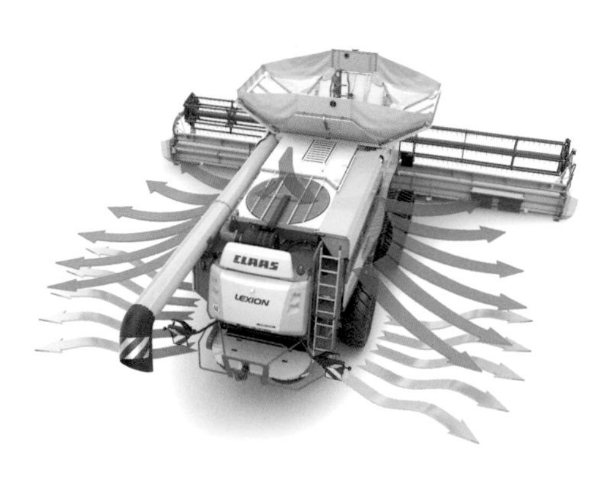

DYNAMIC COOLING saves fuel by controlling the fan speed in accordance with current demand.

"New cultivation methods are becoming increasingly relevant. We position our machines to meet customers' wishes".

KLAUS SCHÄFER

and its "baby brother" the JUNIOR had made the combine harvester concept a viable proposition for farmers in Germany and Europe, thanks in part to the revolutionary cross-longitudinal flow system, the company enjoyed a very strong position in the market. The competition realised that it could not hold its own on this basis and that it would have to come up with something new to counter the CLAAS models. In Harsewinkel, in turn, it was felt that the time for the development of self-propelled machines had come. There could be no question of resting on the successful sales of the SUPER. Instead, the CLAAS developers worked feverishly on the first self-propelled machine, which was launched in 1953 with the name HERCULES. This combine harvester worked on the longitudinal flow principle with all the key elements, such as the hydraulics and the air-cooled four-cylinder engine, being developed and manufactured by the company itself. This machine also owed its success to its versatility: as well as handling the usual grains, the HERCULES could harvest peas, rapeseed, maize, clover and grass seed. In fact, thanks to its versatility, this machine remains to this day the foundation for all subsequent CLAAS developments in the self-propelled sector.

The presentation of the MATADOR in 1962 heralded the era of the large-scale combine harvesters. Here, too, it was the market that called for such machines: fields were getting bigger and in Germany the economic miracle meant that increasing numbers of customers were ready to invest in large-scale combine harvesters. For CLAAS, this was the start of an era in which a large number of new models pushed back the boundaries of optimum performance. SENATOR, CONSUL, DOMINATOR and MEGA: for agricultural machinery enthusiasts, these names from the sixties, seventies and eighties still stand for maximum performance and ever greater efficiency as well as elegance and comfort.

1996 – the first LEXION: the new top-of-the-range class

When CLAAS launched the LEXION 480 in 1996, it was another advance into a new performance class for the company.

With a top speed of 40 km/h, the LEXION can reach the next job location quickly (also applies to wheeled version).

The customers were already familiar with the principle of the APS threshing system from the MEGA model series, but now there was an innovation which represented a landmark in the history of agricultural machinery: a threshing drum diameter of 600 mm. Until then, the drum diameter had been set at 450 mm. A change of size? Little short of unthinkable for some agricultural machinery specialists. Opinion was fiercely divided between those who advocated an enlarged drum and the traditionalists. Arguments were exchanged, concerns raised – but in the end, CLAAS broke with the established order and developed a threshing drum with a diameter of 600 mm. Despite its size, the first LEXION was notable for its excellent user-friendliness and driving comfort. Particular attention had been paid to revising the cab design and equipment: the CEBIS on-board computer handled tasks which could be automated while the LASER PILOT and GPS PILOT steering systems simplified the operator's task. Thomas Barrelmeyer sums up these innovations in the first LEXION generation: "Digitalisation had arrived in the world of threshing technology."

SCAN + WATCH

It was in 2003 that the company brought the next LEXION generation to the market: CLAAS presented the 500 series with the proud suffix "NEXT GENERATION". The innovations in the new model series included electrically adjustable rotor separation areas and the GRAINMETER, which displays the volume and composition of the returns on the CEBIS screen.

Just two years later, CLAAS launched the LEXION 600 as the new top-of-the-range model and the pace of new developments increased yet again. This machine was the world's most powerful combine harvester at the time: in favourable conditions, it was capable of harvesting up to 70 tonnes of wheat per hour. A notable development a few years later was the option of the TERRA TRAC half-track system for low soil compaction which allowed the LEXION 760 to travel on the road at as much as 40 km/h. This meant that the LEXION was now also the fastest combine harvester in the world!

With the new LEXION and the innovations at the very heart of the machine, CLAAS is adding a new chapter to its history of combine harvester development. Anyone who thinks there is scarcely any scope left for revolutions in the field of mechanical engineering is in for a surprise: the technology at the heart of the new LEXION range is more powerful and sustainable than ever.

Digitalisation pioneer in Berlin

In the German capital 365FarmNet is working on an open and neutral platform which will allow farmers to manage their entire farm digitally. Managing director Maximilian von Löbbecke and his team are developing innovative software solutions which they are very confident will offer customers discernible added value.

TEXT Andre Bosse **PHOTOS** Bengt Stiller

To illustrate the capabilities of the app, the developers at 365FarmNet use a series of tiles as a means of visualising its various options and functions.

› SOFTWARE
› NETWORKING
› HARVEST LOGISTICS

It's early morning in Hausvogteiplatz in central Berlin, and the employees of companies whose offices are located at No. 10 are strolling into the foyer and summoning the lift.

Life in this part of the city is not hectic, but somehow bustling and laid-back at the same time. The CLAAS subsidiary 365FarmNet has just leased two floors in this building. More than 80 people work here, mainly as software developers or data and marketing specialists. Maximilian von Löbbecke's office is glazed. He is an approachable boss. The managing director studied mechanical engineering and medical engineering, then worked for Fiat and CNH for ten years where his managerial responsibilities spanned production, development and marketing. He learnt how businesses think. In 2009 Löbbecke changed direction and founded a start-up company based in Silicon Valley, California that developed digital platforms, of which he was CEO.
Then family tradition called him back: the von Löbbeckes are ninth-generation farmers and when he came into contact with CLAAS in 2012 through his business dealings, he was particularly taken with their idea of connecting all areas of farm management using a digital platform: "I immediately saw myself in this field because it required three things which appealed to me: corporate experience, a desire to start a new business and farming in your DNA."

In its early stages the platform was developed at the company's headquarters in Harsewinkel, but CLAAS soon decided to make the subsidiary as autonomous as possible. "You can only manage the delicate balancing act between corporate and start-up culture if the new company is able to operate in a different location to the main business." And there was another good reason for making 365FarmNet an autonomous brand as far as possible: "What we're doing here is designing a neutral platform to connect all machines regardless of the manufacturer. If we are based in Harsewinkel it would be difficult to convey this sense of neutrality. In Berlin, it's much easier for us to do this." In Europe in particular, farmers have a wide range of options for equipping their vehicle and machinery fleet. "Hardly any customers rely on a single brand." And yet they were demanding ever greater machine connectivity. Maximilian von Löbbecke explains why: "As a farmer, I face two challenges today: the first is the rising costs of yields that have remained more or less the same, which means that farmers have to make their processes more efficient to ensure that the business remains profitable. To make that happen, I need to know what's happening on my farm. The second problem is the increasing complexity of statutory regulations, which I can no longer handle with pen and paper."

So what customers urgently need is a user-friendly tool that creates transparency and simplifies documentation. And although software solutions have been available in the agricultural industry for several years, "they tend to be segregated "silo" systems that don't communicate with one another", explains von Löbbecke. The 365FarmNet platform connects these silos. Harvest logistics, herd management, inventory management, crop analysis, weather data and accounts – the platform's modules provide comprehensive management options that customers can tailor to their needs. "Special applications can even predict that your favourite cow will be ill tomorrow because the sensors can detect that the animal's readings are not quite right today."

To make this happen, 365FarmNet cooperates with several partners – some 30 companies are involved in providing digital services, including direct competitors in agricultural engineering. "The fact that this platform opens the door to these partnerships for the benefit of customers – that's the greatest innovation in my view", says von Löbbecke. The software is designed to be used on smartphones and tablets and the data are stored in the cloud. This makes it incredibly simple to use. "Customers don't even realise what a complex tool they're using and just how many applications and connections are working away behind the scenes."

But how does the company make money? Although the basic version of 365FarmNet is free, customers buy additional modules. The cost depends on the size of the farm. The total package including all services is currently just over 80 euros a month for a 100-hectare farm. Customers can get started as soon as they have entered their data in the system. It doesn't take long, explains von Löbbecke, "around two hours, then the platform knows everything it needs to know". But there's also a psychological aspects to this: "Our customers are willing to share their data if they know how it benefits them. The biggest task facing us in marketing is to clearly explain this added value."

Data security is also a particularly sensitive issue. "Customers are entrusting us with their company secrets and naturally it is of paramount importance to them that we keep them safe." With this in mind, the IT experts have constructed a complex safety architecture and the company only uses data centres based in Germany. "In addition, our story has to be convincing", says von Löbbecke. The time is simply ripe for a digital management platform. "And no matter how proud farmers are of their machines: what interests them most at the end of the day is the output they produce." Of course companies like CLAAS will always make money from selling machines, "but in agricultural engineering, just like the car industry, business models which break new ground will become increasingly important in future." Automation, artificial intelligence, subscription models for machines – "Further innovations are on the horizon, all of which will have one thing in common: they will work digitally".

Gamification: the developers of 365FarmNet have been influenced by game-design elements which is why game controllers are used as a tool.

"Our goal: to understand our customer as best we can"

As a partner in the CLAAS Group, Volker Claas is in charge of the Market Research department within the company. In this interview he defines his role, emphasises the value of innovation and explains why it is so difficult to put a price on it.

TEXT André Bosse **PHOTOS** Lukas Kawa

"I think nowadays we are seeing the greatest advances not on the mechanical side, but on the interfaces between electronics and digitalisation".

VOLKER CLAAS

›MARKET RESEARCH
›CUSTOMER NEEDS
›TRUST

Mr Claas, what exactly does CLAAS Market Research involve?
Well, one thing that Market Research cannot do is look into the future. None of us has a crystal ball in which we can see what's going to happen next. What we do is analyse the here-and-now. But we focus less on the market as a whole, and more on the customer. So perhaps it would be more appropriate to talk about customer research: we research customers, we want to know what they need and what they are prepared to pay for it. It's also important to gain insights into our customer groups: who's using our machines, who's interested in them? What do these people do – and above all: why do they buy CLAAS and not another brand? So our goal is to understand the potentially attainable clientele as best we can.

To pick up on your question: why does a customer choose to buy CLAAS, what is their motivation?
Hardly any customers are in a position to compare big machines from different manufacturers. Nobody orders several combine harvesters so that they can test drive them on their field, let alone conduct the tests for an entire year so that they can check for wear. Some consultants provide customers with data from these kind of comparative tests, but our strongest currency is trust. Someone who has had good experience with our machines generally has little reason to change. We are confident enough that we can do justice to the trust placed in us – especially because we perform these tests and know how our machines perform compared with those of our competitors.

From the point of view of market research, why are innovations important for customers?
Because at their best they can meet a need that the customer – as the user of the machine – has identified. Perhaps the old machine didn't do something well enough. Or maybe there was a useful function that it was unable to perform. A good innovation fills this gap – and the customer sees the benefit immediately.

Volker Claas (left) in conversation with a colleague: regular discussions are essential, they help us to incorporate customer needs into the work we do at CLAAS.

So does this make customers themselves drivers of innovation?

In many instances, yes: many customers are keen inventors on the side, and some innovations have found their way to CLAAS via this route, such as the SHREDLAGE® technology, for example. However, competition is by far the greater driving force for innovation. Certainly our customers have wishes which they communicate to a greater or lesser extent and in our customer satisfaction surveys we make a point of giving customers an opportunity to comment on their scores so we can see exactly where the problems lie. Nowadays of course, we can get this feedback much more directly using tools such as marketing platforms and customer surveys or simply by going to the farm and talking to them directly.

Is the digital revolution changing the nature of the customer-company-innovation triangle?

What we actually find nowadays is that we are not incorporating new applications into the machine primarily because they are technically feasible. Increasingly we find ourselves turning to customers to understand what they want, e.g. less fuel consumption, reduced soil compaction or simplified on-farm logistics. We pass on this feedback from our market research to the technical departments together with a note: please take a look at this, it contains information that can help you drive forward innovations.

Is there a single customer type?

Yes and no. Of course each customer is different and they all have individual needs. But our target group is smaller and thus more homogenous than for example a car manufacturer's. So it's the job of Market Research to generate an image of our customer that all departments can understand. Ideally, our marketing experts, product managers, mechanical engineers and software developers should all have the same customer in mind. In fact, Market Research requires colleagues in our department to actually focus on the customer and to always ask themselves: how can I do the best for the people who buy and work with our machines through my work?

You mentioned before gaps that an innovation can fill. In your opinion, what areas today have the greatest potential for innovation?

Over one hundred years ago, a professor in agricultural machinery at Munich University actually said that his professorship could be abolished because with the development of the tractor, everything that mattered in his field had already been invented. Looking back at what has happened since then, I think we can safely say that the man was wrong. Nevertheless, I would say with due caution: I think nowadays we are seeing the greatest advances not on the mechanical side but on the interfaces between electronics and digitalisation, for example in the control and sensor systems or in terms of logistical processes.

This has given rise to far-reaching digital innovations, including the 365FarmNet management system jointly developed by the CLAAS sister company. You know your customer, so what do you think: are they ready for this innovation?

Looking to the future, a development like 365FarmNet is certainly sensible and useful. But customers also have to be able to cope with it in the here-and-now. Because if we overwhelm them at this point, and perhaps even lose them, then it will be difficult to entice them back later. That's why it's important that farm management software like 365FarmNet is not only perceived as innovative, but also that it does not present too much of an obstacle to users. The same applies here too: the customer must see the benefit. And see it immediately.

Are your customers guided by their head or their heart?

How much time have we got? (laughs) There's no straight answer to this, our customer is neither an unequivocal Homo economicus, nor driven entirely by their emotions. They fall somewhere between the two poles, but one thing is important: if we succeed in appealing to the customer's emotions, there should be no cognitive dissonance or conflict afterwards. In other words, it's no good if they regret their decision later. So it's all the more important that we convince our customers not just on an emotional level but also in terms of the machine's performance. The same applies to successful innovations: it's important for us to create a narrative, but in the field they must also live up to the promises we have made. Otherwise the innovation has no more value than a school essay below which the teacher has scribbled "failed to answer the question.

The most important currency at CLAAS is the customer's trust. To retain it, people across all departments in the company strive to understand the buyers and address their needs.

That's an "F" then!

Yes, but let's be clear, this harsh judgement should not result in a fear of failure when it comes to innovations. If you develop ten innovations, you should not assume that they're all going to work. The fact that some innovations don't work is just part of the job. Basically, every business is in the game of betting on the future. In the start-up sector, these bets are extremely risky: an estimated 99% of new start-ups go under but those that make it can grow to become huge and incredibly influential concerns. I believe that in our industry it's important to exercise care and foresight when betting on the future. And market research – when combined with economic data or even geo-data – can help ensure that at CLAAS we make the right decisions.

"Market research is important because it helps us better understand customer needs".

VOLKER CLAAS

Your department is also responsible for selecting suitable prices for the machines. Can you put a price on innovations?

If we were sitting in a university seminar I would say: yes, we would use a mathematical model to calculate the potential customer benefits of the innovation, and then make a recommendation. In practice, it's more complicated because we also have to take account of what our competitors are offering and what generally makes the market tick. So prices are defined by customer benefits and the competitive landscape.

Take for example the new LEXION with its innovative features ...

... customers should be delighted because for the money they pay for the machine, they get exceptional performance and a host of innovations. Some of these are obvious, others less so, but the art lies in presenting the innovations so that the customer understands them and thus also recognises why we have incorporated this new development. We must tell the story of the innovation in such a way that the customer clearly recognises the enhanced benefit for themselves and their work as well as recognising our innovative strength. This engenders trust, which is so important for us. Market Research obviously has a role to play in building this trust because it gives us a better understanding of customer needs. But we are still caught up in a delicate balance between external influences such as the weather, political decisions or market developments. To return to the subject of crystal balls, for this reason alone, one would never be enough – we'd need half a dozen of them.

Volker Claas is a member of the Shareholders' Committee, a position which he took over from his father Reinhold Claas. A business graduate from Paderborn University, he is in charge of the Market Research/Pricing division. He lives with his family in Harsewinkel.

A system for generating ideas

What does a company need to be innovative? Why is creative freedom important, what is the impact of stress and time pressure, what changes as a result of digitalisation? Martin Hawlas is in charge of Corporate Research & Development at CLAAS and Christoph Molitor is director of Technology Management. In this interview they discuss what defines a good culture of innovation, why customers often communicate on an equal footing with the developer and why it can be important just to let your mind wander in front of the television of an evening.

TEXT André Bosse **PHOTOS** Lukas Kawa

Mr Hawlas, Mr Molitor, which aspect tends to get overlooked in discussions about innovations?

› CULTURE OF INNOVATION
› COMMUNICATION
› DIGITALISATION

Martin Hawlas: Most people think of an innovation as a technical breakthrough, something amazing and unprecedented that will light up everyone's eyes. This is very important. But I believe that meaningful innovations can also take place within processes and tools. These innovations are just as important. Because in essence, they lay the foundations that make pioneering technical achievements possible.

Christoph Molitor: I think that it is helpful to expand the term. I was watching television yesterday evening and the programme was rather boring. So I let my mind wander and considered how many innovations a person thinks about each day. I think it would add up to quite a lot because everyone strives to optimise things. It's the same with day-to-day work, where supposedly minor technical problems can give rise to innovative solutions. As humans, we are always in search of improvements that provide a benefit. The same is true of CLAAS.

What does a company need to be innovative?

Hawlas: It's important to have a role model, someone who gives a face to the company's guiding principles. We have had this for decades through the individual family members. Then, as a company we have to give our colleagues the creative freedom they need to facilitate a culture of innovation.

Where are these freedoms within the company?

Hawlas: The task of predevelopment is to come up with ideas for new technologies on the understanding that their commercial viability is verified right from the start. "Free predevelopment", on the other hand, is about generating ideas without having to consider whether the idea will one day pay its way or not. We provide a budget and say: "Do something with it, experiment, and if in the end your experiment doesn't work out, then so be it." Departments like this are important because if every division in a company is bound directly and only by economic expectations, then I have problems developing a good culture of innovation. And I'd go even further to say that without this freedom from economic considerations, our XERION would probably never have come into being. During its development there were definitely some moments when it was not commercially viable. Or rather not yet, as we now know. One thing's for sure though, we learned a great deal.

Molitor: Sometimes these freedoms come about spontaneously. Imagine that a machine has a defective part or there are doubts about its functionality. That's not great, but in an ideal scenario it can result in the customer, dealership and CLAAS engineers looking at the problem together, exchanging ideas and coming up with a solution that will prevent the damage occurring in future. What emerges is not some big innovation that everyone in the company is talking about. And yet these small-scale innovations are incredibly important because they too result in a genuine improvement.

So you could say that problems elicit innovations?

Hawlas: Yes, because they demand solutions – and if the conventional approaches don't work, then we just have to find new ones. One thing has changed of late: as developers we used to be far more knowledgeable than our customers. It's true that farmers have always been interested in technology, but they tended to lack an in-depth understanding of mechanical engineering. Nowadays however, many customers communicate with us on an equal footing, especially when it come to the development of new ideas and business models.

How has this change come about?

Molitor: Digitalisation is an important driver. We all live in a digital world, we use smartphones and organise many things online. And farmers too think deeply about whether and how these digital tools can help them in the field and on the farm. For example, they ask what data the machine can collect during harvesting and how they can use and process these data.

Hawlas: I also think that it is now increasingly rare for innovations to arise from flashes of genius, they are far more likely to be the result of interactions.

How does digitalisation help to organise these interactions?

Hawlas: Digitalisation provides us with a wealth of complex data. The art lies in sorting through this volume of data to extract the information that is important for a process and bring it to the fore. We have to weed carefully without missing anything.

Molitor: The great advantage of digital communication is that it enables a direct exchange between those involved. Imagine that during the development of a machine, a problem has to be solved that requires input from various disciplines. We can create a virtual room – even enriched by virtual reality –, where not only our engineers from different departments but also those of our suppliers can meet, exchange information and find a solution together. This interdisciplinary and cross-company communication is becoming increasingly important and widespread.

To what extent does digitalisation impinge on development? Do you still have meticulous engineers who simply want to optimise the machine?

Hawlas: Yes, they still exist and they are important. But they must be open to digitalisation so that their knowledge meshes with the digital options.

What does that mean in concrete terms?

Molitor: Machine optimisation used to focus on throughput, work quality and of course efficiency. Today it's also about providing the customer with machine performance and harvest data.

For Christoph Molitor (left) and Martin Hawlas, the interdisciplinary exchange of ideas is key to a successful culture of innovation at CLAAS.

The customer wants to use machine data about yields and qualities to further optimise his entire process, from drilling to harvesting. To achieve this, the mechanical engineer has to think about the sensor that will record these data. This calls for dialogue between the departments and it's an important part of our innovation culture that we encourage, and even demand, this level of communication. Always on condition that both divisions – mechanical engineering and the IT world – are seen as equally important. Anyone looking at the news today might well imagine that there was nothing more to tell about the mechanical engineering world and that innovations resulted only from digital applications. But that's not the case and we must always be clear about this.

"Encouraging and demanding the exchange of ideas between mechanical engineering and the IT world is an important part of our culture of innovation".

CHRISTOPH MOLITOR

INNOVATIONS
#01 – #15

In a corporate history spanning more than 100 years, CLAAS has brought a multitude of innovations successfully to market. A glimpse at some of these groundbreaking developments shows what has always set CLAAS apart: its pioneering spirit and technical expertise.

Everything in **full view**

Developed in conjunction with farmers all over Europe, the PANORAMIC cab offers a unique sense of space and visibility.

Good visibility is essential, especially when working with a front loader. CLAAS recognised this need very early on and large, wide transparent sunroofs soon became commonplace on all CLAAS tractors. Further developments ultimately led to the unveiling of the PANORAMIC cab in 2014.

#01

NAME
PANORAMIC
PRODUCT
TRACTORS
YEAR
2014

But there was still a very long way to go before bench style seats in utility vehicles would be replaced by closed, air-conditioned cabs. In the early stages of their development, self-propelled harvesting machines and tractors were not equipped with cabs because comfort was simply not part of the brief when it came to designing driver's platforms. A roof – mostly just a canopy to protect against rain and strong sunshine – was long deemed to be sufficient. However, as machines became more powerful, the strain on the driver increased. Sitting on the open platform of a combine harvester, on many days you could only bear the high levels of dust generated during threshing by covering your mouth and nose with a damp cloth. The same was true of tractors. Eventually some manufacturers, including CLAAS, began to offer closed cabs. These made it possible to avoid the worst of the dust.

Over time, more features were added to enhance driver comfort and safety was further improved. Conventional glass roofs were no longer adequate to comply with the FOPS (Falling Object Protective Structure) standard. However, the CLAAS engineers were not content with the simple solution of fitting a metal grille over the glass. So they designed a polycarbonate windscreen which reliably protected the driver if something fell onto the roof while they were at work. With each new cab design, the engineers reduced the width of the crossbeam between the front windscreen and the roof – continuously improving visibility. At the end of the development phase CLAAS unveiled an innovative six-pillar cab: discussions with customers throughout Europe had revealed that farmers wanted a cab with significantly better visibility at the front for front loader work. And that's exactly what the company delivered.

The PANORAMIC cab unveiled in 2014 saw the crossbeam disappear altogether. The glass front windscreen and polycarbonate roof were seamlessly joined to give a 90° vertical field of vision from the cab, while the extremely slender pillars guaranteed an exceptionally good 360° all-round view. This development was well-received by both customers and industry professionals alike – in 2015 the CLAAS PANORAMIC cab received the Gold SIMA Innovations Award.

SCAN + WATCH

Throughput monitor on the DOMINATOR 96

A watchful **eye**

As with cars, optional extras were available for harvesting machines from an early stage. One example was the electronic throughput monitor developed for the DOMINATOR combine harvester series, which informed the driver about grain losses. The main attraction of this device was the indicator in the cab, which showed the driver whether he was threshing at the best forward speed for the DOMINATOR to keep grain losses to an acceptable level.

#02

NAME
THROUGHPUT MONITOR
PRODUCT
COMBINE HARVESTERS
YEAR
1974

The basic principle of the throughput monitor is as follows: grains that were thrown out by the sieve pan rather than landing in the grain tank fell onto sensors fitted with microphones. The signals they produced were amplified and then used to control the display of lights in the driver's cab. The sensors were installed not just in one place but across the entire width of the inside of the machine. This ensured that grain losses were reliably indicated even when the load was unevenly distributed or when driving uphill. Now, grain characteristics vary depending on the type of cereal: the grain size and shape, awns and leaf parts are different. The moisture content also influences their behaviour. The driver could take into account all these parameters using two rotary switches each with eight sensitivity settings. One of the switches changed the sieve pan settings, the other the walker settings. A table was available showing the basic settings for the different grain types.

Despite this handy table, the farmer's working day still began with a specific routine: he would start combine harvesting at "an appropriate forward speed for the crop", as advised in the instructions. After a period of time, if the fourth or fifth lamp on the light strip was constantly illuminated, he would have to switch off the machine, step down, walk to the rear and check grain losses in the straw swath. Then readjust the basic settings accordingly. The development of this helpful device obviously called for an accurate understanding of how the threshed grains behaved inside the combine harvester, knowledge that was readily available at CLAAS.

The innovative throughput monitor was only the start of a wave of automation that was to become a feature of CLAAS designs. The company subsequently integrated the technology into a "harvesting information system", a kind of instrument panel which monitored different combine harvester functions and warned of speed drops affecting key drive elements. A "driver information system" beneath the steering wheel indicated the oil level, cooling water temperature and other metrics and last but not least, an "on-board information system" – also referred to as a fieldwork computer – displayed various performance data and reminded the driver of service intervals.

SCAN + WATCH

CLAAS
ROLLANT 85

ROLLANT
Steel for straw

Ingeniously aligned steel rollers ensured that the ROLLANT baler was ahead of the competition when it came to quality, performance and efficiency. The baler offered customers a range of innovative features and equipment to suit the varied needs of farmers.

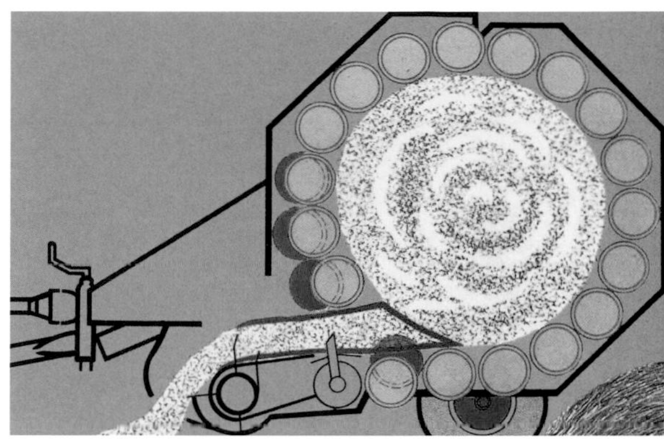

The spiral arrangement of rollers ensured maximum bale density.

#03

NAME
ROLLANT
PRODUCT
BALERS
YEAR
1976

The ROLLANT round baler was introduced by CLAAS in 1976. As the trailed baler travelled through the swath, its rotating tines picked up the hay or straw arranged in long windrows, then a rotary feed scooped the crop into the bale chamber. The walls of this barrel-like chamber were lined right up to the filling port with rotating steel rollers, rather than the customary belts. The chain-driven rollers rotated in the same direction to form the harvested crop into round bales. When the preset baling pressure was reached, the driver simply pressed a button and the machine automatically wrapped several lengths of twine around the bale. Then the rear of the baler opened and the round bale rolled out onto the field.
A seemingly simple process, yet CLAAS has continuously striven to improve the underlying technology. When the company made a landmark decision to switch entirely to fixed chamber balers, CLAAS engineers took a closer look at the way this type of baler worked. It was clear to everyone involved that the arrangement of the rotating steel rollers was key. The crucial innovation introduced in the 1976 machine was to replace the circular arrangement of rollers used by some manufacturers with a spiral layout. With this new design, the spiral ran round a virtually central axis inside the baler.
It soon became apparent that this roller arrangement offered several advantages: the harvested crop was more tightly compacted, producing the firm round bales required for easy storage and handling. The spiral arrangement also helped ensure the smooth running of the baler by pushing the harvested crop in the right direction. Another plus point was that the new baler provided reliable bale rotation, making it easier to deal with variations in moisture content, for example.
CLAAS went on to incorporate a range of other patented innovations into the ROLLANT baler. Among these were the pick-up, originally developed for the baler of the same name, and the CLAAS ROLLATEX net wrapping system. The introduction of MPS, which stands for maximum pressure system, marked a major improvement: the engineers suspended three rollers from the top of the bale chamber on an articulated pivoting segment. This articulated component penetrated into the roller chamber as baling began, which speeded up the start of the bale rotation and compaction process as well as significantly increasing bale density. The success of the ROLLANT baler on the national and international market was based on this wealth of features – but also on the fact that a machine was available to suit any size of farm, with bale weights ranging from 100 kilograms to a tonne depending on the model.

SCAN + WATCH

It all started with **the knotter**

Anyone who has ever wrapped a present knows how difficult it can be to tie a tight knot. CLAAS succeeded in perfecting this process by purely mechanical means almost one hundred years ago. The knotter was the company's first pioneering achievement – and one of the most influential innovations in the agricultural industry.

#04

NAME
KNOTTER
PRODUCT
STRAW BINDERS
YEAR
1921

Patented in 1921 under patent number 372140, the CLAAS knotter was the Claas brothers' very first patented product and the solution to a perennial problem. Straw binders had been on the market since the end of the 19th century, and all had the same flaw: the mechanical knotting devices did not produce secure knots, many of them came undone, causing the bundles of straw to fall apart. The CLAAS knotter solved this problem by using a mechanism that tied a firm knot, no matter how thick the twine. How does the knotter work? The process goes something like this; it unwinds a length of twine from the roll and keeping it tightly tensioned, wraps the twine round the straw bundle and back over itself, then keeping both strands tightly secured, ties a knot and cuts the yarn – before returning to the starting position ready for the next knot. It repeats this cycle over and over again. The unique feature of the CLAAS knotter was, and still is, the floating upper jaw on the bill hook. It opens and closes whilst simultaneously turning once on its own axis to form a loop. The movement of the straw bale through the baler generates traction which helps to produce a tight loop knot. At the same time, the retaining plate or clamping disc holds the two twines placed one on top of the other securely together until the knot is tied and the twine cut. The bill hook with floating jaw was such a successful symbol that the knotter went on to become the hallmark of CLAAS. There is no doubt that the knotter is a true once-in-a-century invention. Two more patents were to follow, making the brothers' original invention a perennial success which has been continuously enhanced and is still used in CLAAS balers today. The improvements include adapting the knotter mechanism to cater for ever higher output rates, bale pressures and changing conditions of use. The twine retaining wheel, for example, has been given two additional deflecting lugs to increase the twine holding force. The designers also made the bill hook larger to accommodate thick plastic twine up to 100 linear m/kg, and most knotter parts are now precision-cast. This development is largely responsible for the increased strength and wear tolerance of the updated knotter design.

Knotters have been manufactured at the CLAAS baler factory in Metz, France since 1981. Here, the workforce currently produces some 7000 knotters per year. These key components guarantee the smooth functioning of the square balers also produced at the factory.

Incidentally, for some time now a robot has taken on the task of manufacturing the knotter frame. The freely programmable production assistant uses a wide range of machine tools. With the aid of integrated inspection systems, it can even check the quality of the work it undertakes. An almost 100-year-old invention enhanced by robotics: there can be no better illustration of the interplay between tradition and innovation that is such a feature of CLAAS!

SCAN + WATCH

Quality control: the perfect knot requires a great deal of precision

The APC offers drivers maximum convenience combined with consistently high bale density.

AUTOMATIC PRESSURE CONTROL
The right pressure at all times

Optimum baling pressure at the push of a button: automatic pressure control (APC) ensures that QUADRANT balers achieve optimum output at all times. This innovative system relies on sensors which are fitted to the optimised version of the classic CLAAS knotter.

#05

NAME
AUTOMATIC PRESSURE CONTROL
PRODUCT
BALERS
YEAR
2016

CLAAS QUADRANT square balers are powerful machines with outstanding performance potential. They can produce bales weighing over 500 kg each, depending on the material being baled and the bale length. Square bales are very easy to store because of their shape. Gentle handling of the harvested fodder and consistent bale quality are key to successful baling. At the same time, farmers want to make the most of the narrow harvest window. This can be achieved if the QUADRANT driver pushes the machine right up to its limit. A new feature called APC introduced in 2016 ensures that balers can safely be operated at their maximum output.

APC stands for AUTOMATIC PRESSURE CONTROL. This system is particularly helpful for inexperienced baler drivers, while for the company, it is a genuinely unique selling point: APC could only be CLAAS! The technology ensures that the QUADRANT baler is always driven at the limits of its performance whilst eliminating the risk of overloading. To do so, the driver simply enters the desired bale density and the quality of binding twine being used in the ISOBUS terminal before starting the baling process. The APC takes charge of the rest. Sensors on the mainframe and on the knotter continually measure the forces generated and the system automatically adjusts the degree of compaction of the square bales accordingly.

In addition to the innovative APC, the new QUADRANT balers also feature the company's oldest ever patent: the knotter, first patented by CLAAS in 1921 and significantly improved since then. The square balers have six of the latest knotters. Three of them are fitted with sensors which continually transmit relevant metrics to the APC. The integrated weighing system has four sensors mounted on the bale ramp which also measure the weight of the finished bale. These data are not only of interest to contractors who calculate their output per tonne. The automatically produced weighing data are also useful for planning purposes and in future it will possible to relay them to the office or to a farm management software system using CLAAS TELEMATICS.

SCAN + WATCH

EN-GJS-500-7
0854 712.0

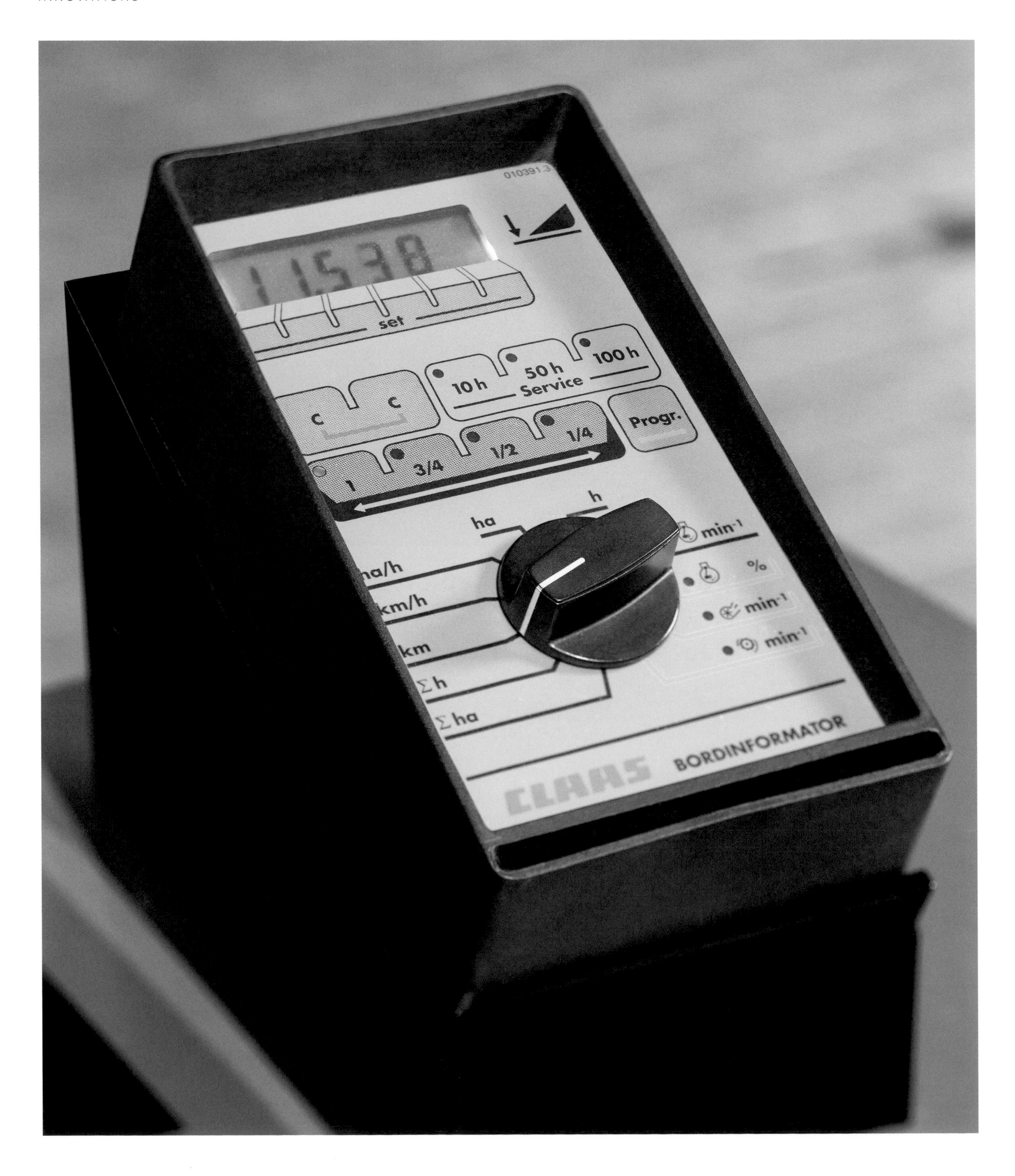
010391.3
set
10 h
50 h
100 h
Service
C
C
Progr.
1
3/4
1/2
1/4
h
ha
min-1
%
ha/h
km/h
km
min-1
min-1
Σh
Σha
CLAAS
BORDINFORMATOR

The first **digital display**

CLAAS introduced the first electronic "fieldwork computer" for certain models in the DOMINATOR and COMMANDOR combine harvester series in 1985, marking an important early step towards digitalisation.

Switches and lights: in the 1980s, the cab began to look more and more like a cockpit.

#06

NAME
FIELDWORK COMPUTER
PRODUCT
COMBINE HARVESTERS
YEAR
1985

At a time when pocket calculators could do more and more and the first PCs and games consoles began to make their way into our homes, CLAAS developed a device equipped with sensors and a microprocessor which provided the combine harvester operator with useful information. Even then, without GPS the small device was able to calculate the part and total area worked and provide information about the number of working hours spent on individual areas. This information was of particular interest to subcontractors as they now had access to an electronically generated data set for invoicing purposes which gave a clear indication of the number of hours spent on specific areas.

The device was easy-to-use. It had a dust- and moisture-resistant keypad and a rotary switch on which the driver could preselect the desired function. When the machine was running, the fieldwork computer also calculated the speed of up to three different shafts, each of which was fitted with sensors. The system even reminded the driver of specific service intervals – information that was particularly relevant for combine harvesters since individual units in the machine had different service intervals. If these were missed, faults could occur or the machine could break down completely – a scenario that farmers wanted to prevent at all costs, given the narrow harvest window.

SCAN + WATCH

The fieldwork computer enabled CLAAS to gain vital experience in the electronics sector which it then applied to subsequent developments. In the mid-80s the pace of development in this sector rapidly accelerated in a market that was receptive to innovation – and CLAAS was perfectly positioned to quickly become one of the dominant players in the field.

PICK-UP

Better hay and straw handling with the **CLAAS PICK UP**

The PICK UP baler marked a major step forward in mechanised farming: CLAAS promoted this true all-rounder with an effective advertising and marketing campaign, illustrating very early on the company's keen instinct for knowing just what customers needed.

#07

NAME
PICK UP
PRODUCT
BALERS
YEAR
1934

Gathering straw or hay from the field with a pitchfork or other tool was a strenuous and labour-intensive task. Many workers were required, but farm labour was in short supply even in the 1930s. The CLAAS PICK UP baler mechanised all the individual stages – from gathering and baling to loading. The PICK UP either deposited bales in large groups on the field for workers to bring in later or loaded them directly onto the wagon via a bale chute.

The separate elements – pick-up rake, intermediate feed, baler and loading system – were arranged one behind the other in the PICK UP. Its compact, lightweight design made it possible for the baler to be trailed to one side of the tractor and driven by power take-off shaft without pulling the tractor off-track. The great advantage of this design was that since the mown crop was not in the tractor's path, the tractor avoided driving over and flattening the swath.

Shortly after this innovation, CLAAS launched its first combine harvester, the Mower-Thresher-Binder (MDB), which performed all the harvesting steps including straw binding. But there was still a good argument for dividing the work between the combine harvester and the PICK UP: on the one hand, it allowed straw gathering and threshing to be carried out independently, which was very useful, especially during dry spells. At the same time, efficiency gains could be achieved by using two specialised machines. CLAAS was one of the first companies in the world to offer this type of trailed, PTO-driven pick-up baler. All it needed was a 25 hp tractor. Before taking off in Germany, the PICK UP enjoyed initial success abroad, especially in Great Britain, where from 1934 onwards it was much in demand.

The quality of fodder that could be achieved with the PICK UP was of particularly interest to dairy farmers – a fact that the CLAAS advertising department used to good effect. The PICK UP brochure featured a cow with the following words coming out of its mouth: "Give me tasty "PICK-UP" fodder and I'll give you lots of the creamiest butter." The slogan proved a hit; from 1937 the PICK-UP was to become a bestseller in Germany despite commanding a price of 1850 Reichsmark. The patented pick-up baler continued to enjoy export success until the outbreak of the Second World War, with machines going to France, Holland, Denmark, Sweden and Romania.

SCAN + WATCH

Gathering, baling and loading in a single operation greatly simplified the task of bringing in the hay and straw.

Cleaning in **three dimensions**

Conventional sieve pan cleaning systems reach their limits on sloping sites. The steeper the slope, the greater the risk of a mat forming in the sieve pan, resulting in a drop in performance of the entire machine. In 1983 CLAAS designers introduced the 3D cleaning system to address this issue.

#08

NAME
3D CLEANING SYSTEM
PRODUCT
COMBINE HARVESTERS
YEAR
1983

The problem with the conventional system was that when the machine tipped sideways, the crop in the sieve pan slid to the downhill side of the pan, reducing the cleaning effect and increasing grain losses. To avoid exceeding the acceptable loss rate of one percent, the driver would have to reduce the ground speed of the combine harvester. Even a ten percent incline reduced performance by up to a quarter, while a twenty percent incline cut it by as much as a half.

Initial attempts to incorporate side rakes into the straw walkers which flung the crop back to the middle of the sieve pan when the combine harvester was operating on a slope proved impractical. The rakes quickly became clogged and then jammed.

The solution that CLAAS designers came up with was an automatic control circuit which actively controlled the cleaning system in the combine harvester. The advantage of this was that it could be incorporated into the combine harvester design with comparatively little effort. The only additional element that had to be installed was a hydraulic control unit consisting an oil-damped gravity pendulum and a control valve. The device automatically detected the direction and intensity of lateral oscillations in the upper sieve with the aid of a pivot arm. Modified sieve bearings and suspension points provided the necessary flexibility: the pendulum matched its deflection to the angle of slope of the combine harvester and the unevenness of the ground and transmitted control signals to the system controlling the movement of the upper sieve. So as well as moving from side to side and backward and forward, the upper sieve additionally oscillated up and down in the third dimension to counteract the effect of working on a slope.

From 1983 onwards, not only did the company offer the new CLAAS 3D cleaning system on all new combine harvester models, it also supplied retrofit kits for older machines. Customers were amazed by the effectiveness of the new dynamic slope levelling system. The output of the combine harvester remained constant even on slopes of up to twenty percent, where it performed just as effectively as it did on level ground.

SCAN + WATCH

Even on steep slopes, the crop is evenly distributed across the upper sieve

ROLLATEX
Speed wrapping

Unbeatable efficiency, inspired by nature: ROLLATEX net wrapping superseded the old twine tying process and saved farmers a great deal of time.

Record speed: ROLLATEX net wrapping reduced the wrapping process to ten seconds, boosting productivity by up to fifty percent.

#09

NAME
ROLLATEX
PRODUCT
BALERS
YEAR
1983

The hook-and-loop Velcro-type fastener is a wonderful invention inspired by nature. The burs of the burdock are covered in tiny barbed hooks which stick to the fur of passing animals. This is how the plant disperses its seeds. Swiss engineer Georges de Mestral observed this principle after repeatedly removing burs entangled in his dog's coat. When he examined them under the microscope, he immediately realised that nature had designed a fastening technology which held fast but could be undone without damaging the connected parts.

When designers were seeking to improve the working speed of the ROLLANT round baler, it was admittedly not their original intention to replace the existing twine tying system with hook-and-loop fasteners. But the principle borrowed from nature provided a template for the CLAAS engineers to draw on during their deliberations. They developed a process whereby a wide-meshed net wrap could be unrolled into the top of the baler and wrapped around the bale as it rotated in the chamber. The system also included a cutting unit which cut the net wrap to a predetermined length.

After several trials, the developers decided on a wide-meshed polyethylene (PE) net wrap. This wrap had a certain amount of give widthways, but was very dimensionally stable lengthways and capable of hooking into virtually any baled material. As a result, no additional fastening mechanism was required. One-and-a-half wraps around a round bale was enough to secure it firmly and retain its shape – which was particularly important when it came to storage. And just like conventional hook-and-loop fasteners, the net could be removed without applying excessive force leaving no residue and furthermore it could be recycled – PE is 100 percent recyclable. The main advantage of the new process, however, was that it was considerably faster than the twine tying method which had been used thus far: thanks to the innovative net wrapping technology, it now took just ten to fifteen seconds to wrap and set down a round bale, a vast improvement compared with the old method which took three to four times as long.

SCAN + WATCH

MÄH-DRESCH-BINDER
CLAAS
CLAAS

Threshing **against the grain**

The CLAAS Mower-Thresher-Binder revolutionised the European grain harvest.

CLAAS did not invent the combine harvester, but it did adapt this new technology to German and European harvesting conditions, overcoming enormous resistance and major misgivings in the process. Sometimes being innovative means you just have to tough it out.

#10

NAME
MOWER-THRESHER-BINDER
PRODUCT
COMBINE HARVESTERS
YEAR
1936

From around 1900 onwards, combine harvesters were gaining ground in the US. The new machines found favour with customers in the United States and other areas with large arable fields and a fairly dry climate, leading to a five figure increase in the number of units sold per year. Even in those early days of combine harvester development, the machines deployed cutting widths of ten to twelve metres, producing impressive harvesting rates on large fields and boosting the profitability of the farms that used them.

But farmers in Germany had their doubts about this American technology. And not without reason, since the newfangled giants struggled to cope with harvesting conditions in Europe, which were completely different to those in the US – with crop height, large volumes of straw and higher moisture levels proving particularly problematic. Rye, which has stalks up to two metres tall, was not grown in the US. Furthermore, the US machines were simply too large for most German farms. Consequently, several imported combine harvesters performed poorly in harvesting trials, even on large farms – reservations about the new technology appeared to be justified.

Rejecting the majority view and defending your own position even in the face of resistance is not for the faint-hearted. But Karl Vormfelde, an agronomist from Bonn, whose assistant at the time was Walter Gustav Brenner – a gifted designer and inventor – and the Claas brothers were convinced of the benefits of this technology. Together, they looked for ways of adapting the combine harvester for the German and other European markets to make it profitable for farmers. It was to be a long and arduous journey: when the prototype combine harvester failed to convince industry professionals at a demonstration in 1932 and other agricultural machinery manufacturers dismissed it out of hand, August Claas, undeterred, responded by saying "if the others don't want to get involved, we'll do it on our own!"

The thresher trio Vormfelde, Brenner and Claas radically overhauled the design principle and came up with a trailed system consisting largely of tried and tested components. Four years later, the new mower-thresher-binder – dubbed MDB from its German name – was ready for series production. A successful demonstration on the Zschernitz estate in Saxony was to go down in the annals of agricultural machinery history, tongue in cheek, as the "Victory of Zschernitz". The farming industry was convinced by the machine's high harvesting output and its price-performance ratio.

Series production began in 1937 and CLAAS grew rapidly. When it came to marketing the machine, the shortage of agricultural labour at the time was turned to an advantage. The CLAAS argument that the MDB not only boosted efficiency but also reduced the number of labourers required proved persuasive.

SCAN + WATCH

Electro-optical sensors use pulses of light to scan between the standing crop and the stubble and precision-guide the combine harvester automatically right up to the edge of the crop.

Right up to the **edge**

The LASER PILOT sensor system makes harvesting operations faster and more efficient because it relieves the operator of the burden of steering and always guides the CLAAS LEXION combine harvester right to the edge of the crop. Introduced in 1999, this innovation not only makes the driver's job easier, it also increased output.

#11

NAME
LASER PILOT
PRODUCT
COMBINE HARVESTERS
YEAR
1999

Prior to the introduction of the LASER PILOT, combine harvester drivers spent around 60 percent of their time adjusting the steering to keep the machine on track. This innovation allowed them to give the machine settings their full attention instead. And it paid dividends: thanks to LASER PILOT, harvesting operations were completed in less time and output increased.

The developers wanted to ensure that the driver could easily override the system. So as soon as the driver touched the steering wheel, for example to initiate a turning manoeuvre, the LASER PILOT disengaged. To reactivate it, the driver simply had to depress the foot switch. The maintenance-free eye-like optical sensor system was mounted on a mast on the left side of the cutter bar. It was somewhat reminiscent of the head of the friendly robot in the animated film Wall-E. While he was tasked with cleaning up an entire planet, the job of the LASER PILOT was to perform specific tasks more rapidly, and above all, more accurately with the aid of intelligent sensor systems.

After coming up with the idea of this type of guidance system, CLAAS developers spent five years designing the sensor system, associated electronics and software before the product was ready for series production. The basic idea was to use sensors to detect the easily discernible difference between the stubble and the unharvested standing crop and use the data obtained to guide the steering automatically. In other words, the edge of the crop was the limiting factor. This was done by projecting harmless laser beams invisible to the human eye from the left side of the cutter bar at a 12° angle several metres ahead and slightly downwards. These measuring pulses were reflected back and then analysed by the software. The slight difference in time delay between the optical echoes from the stubble and those from the standing crop made this possible. On the basis of this difference, the system calculated a correction signal which it then compared with the steering signals from the combine harvester. The electronic system used these data to generate the perfect track and guide the machine accurately along the edge of the crop.

SCAN + WATCH

claas
claas
HARSEWINKEL i.WESTF.

SUPER **Transverse and longitudinal**

Soon after the end of the Second World War, CLAAS launched a combine harvester that was to dominate the combine harvester market in Germany and Europe for years to come. The enormous success of the SUPER line was due to its unrivalled threshing quality based on an innovative transverse and longitudinal flow process.

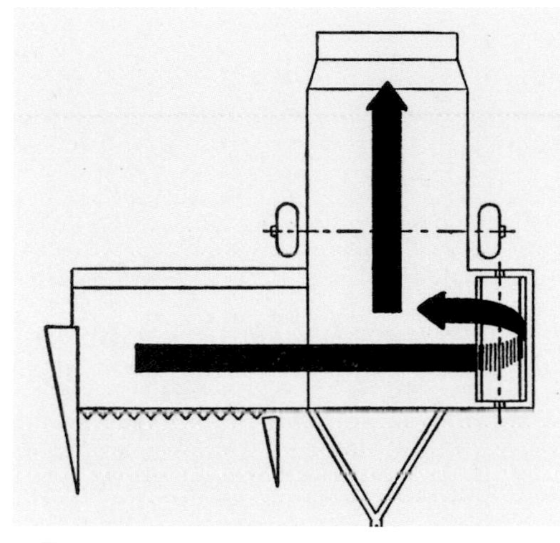

The transverse and longitudinal flow process: one of 197 patents filed by CLAAS at the time.

#12

NAME
SUPER
PRODUCT
COMBINE HARVESTERS
YEAR
1946

Before CLAAS introduced the SUPER line in 1946, there were only two types of combine harvester on the market: longitudinal or transverse flow machines. The former were similar to self-propelled machines in that the crop was conveyed through the machine in a straight line. But the problem with the trailed versions of this design was that the machine was towed alongside the tractor, which reduced the driver's ability to control the machine combination, especially on difficult terrain. Trailed transverse flow combine harvesters were equally difficult to handle on account of their width. In addition, the vibrating sieves used to separate the residual grain – known as straw walkers – were rather short, so the yields were not always satisfactory. On top of that, farmers would often have to manually mow round the edge of the field initially so that the combine harvester combination could get onto the field without cutting a track through the harvest-ready crop.
Discussions about the pros and cons of combine harvesters compared with stationary threshing machines were still far from over, especially in Germany. But it was becoming clear that the combine harvester was increasingly finding favour among farmers. Self-propelled machines were available, but they were extremely expensive. So most customers invested in a trailed model – driven by PTO shaft or auxiliary motor – to bring in the harvest.

SCAN + WATCH

CLAAS eventually succeeded in combining the benefits of longitudinal and transverse systems in the SUPER while avoiding the pitfalls of both. The machine itself was arranged in a straight line behind the tractor, which made it very manoeuvrable. The cutter bar table and reel were the only components that protruded sideways outside the tractor track during harvesting. The engineers fitted the feed rake – a device that conveyed the cut crop to the threshing drum – on the inside of the machine to make the design compact. In transport mode with reel detached and mower folded-up, the SUPER was barely wider than a tractor.
What made the design of the SUPER special, however, was the conical surfaces of the threshing drum. When the threshed ears were thrown against the sides of the drum, the stems rotated by 90 degrees so that they landed crossways on the walker before being conveyed to the straw press. This technology improved residual grain separation in the straw walkers and also created the perfect conditions for loading the straw press. Blockages resulting in interruptions to harvesting operations were a rare occurrence with the SUPER.

CLAAS
40
CROP TIGER
CLAAS
C 320

TAF
Technology for India

The TANGENTIAL AXIAL FLOW (or TAF for short) threshing system is extremely versatile. Machines fitted with this system are suitable for a wide range of field crops and deliver clean, undamaged grain even in wet conditions or when harvesting green crops.

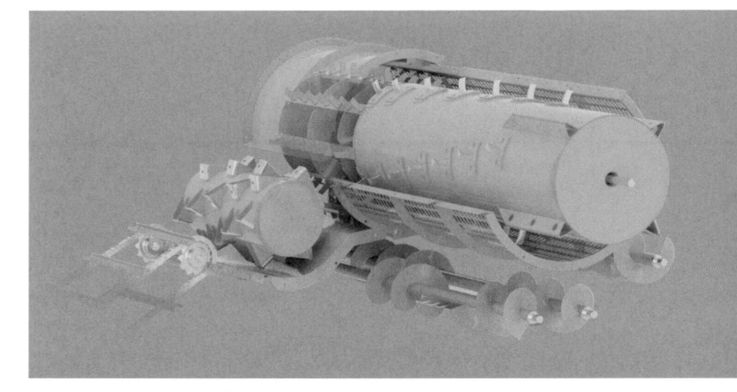

TAF threshing system: rice rotor specially developed for difficult field conditions.

#13

NAME
TAF
PRODUCT
COMBINE HARVESTERS
YEAR
1993

The TAF basically consists of one or two threshing drums arranged crossways to the direction of travel, while residual grain separation takes place in the axle direction. This design has the advantage of being extremely robust – a particularly important factor when harvesting rice, as rice plants accumulate silicon crystals which increase wear on harvesting machines.

CLAAS had already gained experience of rice harvesters in European countries, because rice is grown in Italy, Spain and France. However, India is the largest market for rice combine harvesters. By the end of the 1960s, the company had already set its sights on this market. Contacts with the Indian company Escorts Ltd. resulted in the export of DOMINATOR combine harvesters initially, and by the end of the 1980s to the formation of a joint venture based in Faridabad, a city in northern India. Early blueprints for a "Made in India" machine closely resembled the design which would subsequently be marketed with great success under the name CROP TIGER.

Small by European standards and weighing three and half tonnes, this combine harvester was fitted with rubber tracks and a TAF threshing unit. Series production of the CROP TIGER TERRA TRAC began in Faridabad in 1993 and just one year later the factory received a large order for 27 CROP TIGER for the South Korean market. This country was soon joined by Taiwan to become increasingly important export markets for the Indian site and by 1999 the 1000th "Made in India" CROP TIGER had rolled off the production line, half of which were destined for export. The joint venture was taken over by CLAAS in 2002 and rebranded as CLAAS India in 2003. The company then opened a new factory in Chandigarh in the Indian state of Punjab. Further sites soon followed. Various versions of the CLAAS CROP TIGER with TAF threshing unit are now available, including ones with wheels. The series played an important role in mechanising agriculture in India and other Asian countries and also made it possible for contract farming to emerge as a profession in these regions. CLAAS India celebrated production of the 8000th CROP TIGER in 2016. Other products developed specifically for the Asian market include the PADDY PANTHER rice planter, the MARKANT baler, the DISCO and CORTO mowers and the LINER swather.

SCAN + WATCH

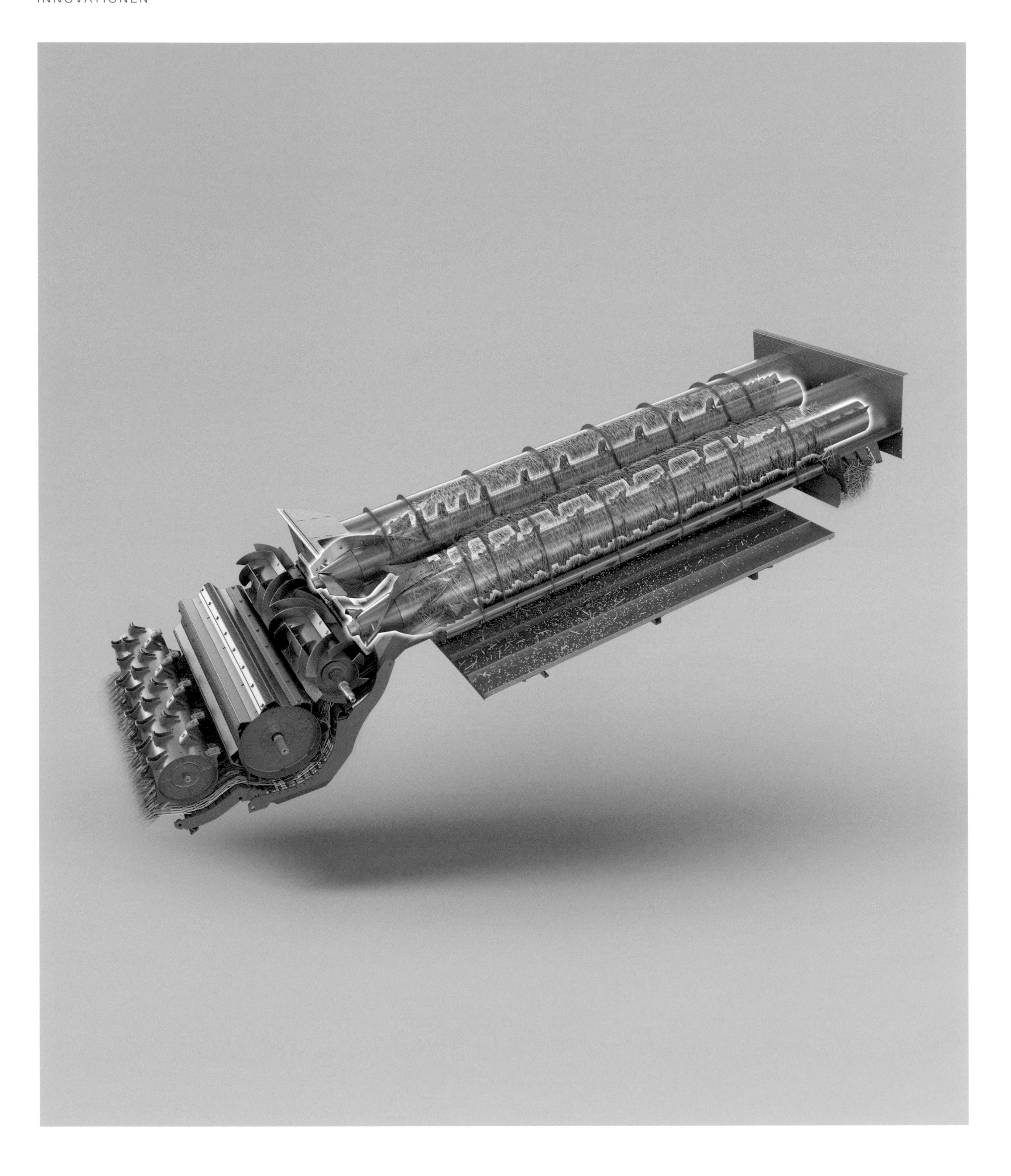

APS HYBRID
The perfect combination

The APS HYBRID system took combine harvesting efficiency to a completely new level. Launched in 1995 and introduced with the LEXION 480 in 1996, this innovation combined two technologies in one application.

Top-of-the-range LEXION: the APS HYBRID SYSTEM is still the benchmark when it comes to combine harvesting technology.

#14

NAME
APS HYBRID SYSTEM
PRODUCT
COMBINE HARVESTERS
YEAR
1995

A combine harvester has two ways of collecting the grain inside the cut ears. In the first stage, the threshing drum in the threshing unit extracts around 90 percent of the grains from the freshly harvested plant parts. Then, depending on the type of combine harvester, any remaining grain is separated from the straw either by straw walkers or rotary cylinders (also referred to as separator rotors). The use of straw walkers for residual grain separation in combine harvesters had long been a thorny issue. The introduction of the innovative CLAAS APS HYBRID was a game changer.

APS stands for Acceleration & Pre-Separation. The aim of the system was to thresh the grains from the ears as thoroughly as possible without damaging them, thus ensuring that the straw produced was also of good quality. Logically, you might think this could be achieved by increasing the rotational speed of the pre-acceleration and so increasing throughput. But the problem with this is that if too much material enters the threshing drum due to pre-acceleration, the grain losses increase and the machine loses efficiency.

The solution was simple: a bigger threshing drum. Engineers had settled on a diameter of 450 millimeters for the threshing drum of CLAAS combine harvesters based on years of experience and this was thought to be set in stone. When developing the APS system, CLAAS experts began to question the size – and they came up with a new answer: the APS has a considerably larger threshing drum, 600 millimeters in diameter instead of the usual 450 millimeters and 1700 millimeters wide. This ensures that the pre-accelerated crop is much more evenly distributed when it enters the threshing drum, which improves the threshing performance.

The system is rightly called "HYBRID", because it combines the APS threshing unit originally developed for the CLAAS MEGA combine harvesters with the ROTO-PLUS separation rotors for residual grain separation. Two rotary cylinders installed longitudinally use centrifugal forces to extract the last of the grain from the crop and still provide good straw quality. It is the combination of these two technologies that makes the APS HYBRID system unique. And the innovation has proved its worth: throughput increased by around 20 percent with the introduction of the LEXION without compromising quality or increasing fuel consumption.

SCAN + WATCH

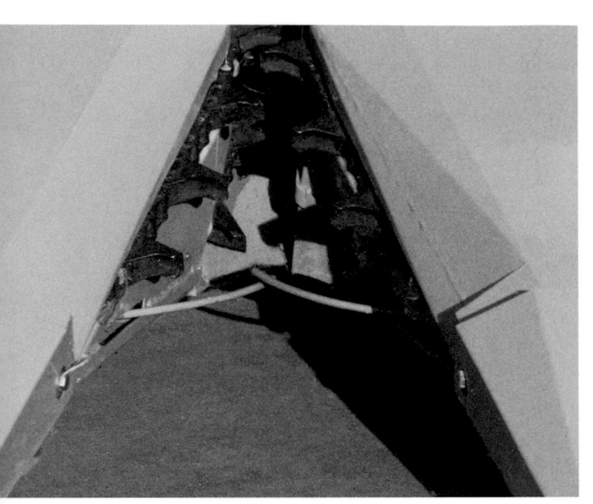

1975 automated steering system: making light work of the maize crop.

The CLAAS **automatic steering system**

Harvesters are such mighty machines that you'd be forgiven for thinking that they need to be handled with brute force. But in fact, a light touch is needed. What modern farm machinery offers nowadays in the way of control systems, satellite navigation and artificial intelligence would have belonged in the realms of science fiction in the 1970s. And so sensors that determined the position of a crop row and sent signals to the steering unit to guide the harvester through the crop were regarded as a groundbreaking, highly innovative step towards automation.

#15

NAME
AUTOMATIC STEERING SYSTEM
PRODUCT
FORAGE HARVESTERS
YEAR
1975

The idea was disarmingly simple: if you look down on the maize header of a forage harvester from above, you can see the conical tines between which the rows of maize plants are lifted and drawn into the forage harvester. The automatic steering system patented in 1977 saw the addition of two slightly curved, movable brackets which scanned the crop row as it was being picked up by the header on the forage harvester. If the direction of travel corresponded to the alignment of the row, no steering signal was sent. But if it changed, the sensors measured the degree of deviation and relayed it to the steering control unit which in turn sent correction signals to the hydraulic steering system. This ensured that the forage harvester automatically steered along the crop rows. Now the driver could let go of the steering wheel and concentrate on adjusting the ground speed as necessary and checking that the chaser bins were being correctly loaded. At the end of the row all the driver had to do was take hold of the steering wheel to disengage the automatic steering system and perform a turning manoeuvre. As he approached the next row, he simply had to depress the foot pedal to re-engage the steering system.

Developed in the 1970s, the engineers' ingenious invention was a welcome improvement that not only made the operator's task easier, it also increased the economic performance of forage harvesters. The automatic steering system proved a godsend on long harvesting days when sooner or later everyone's concentration flags. It never got tired and performed just as well even at dusk or in the dark. So not only was the new CLAAS automatic steering system greatly appreciated by the drivers because it made their job easier, it also brought in a better harvest. As a result, investment in the new technology quickly paid off.

JAGUAR forage harvesters were not the only machines to deploy the automated steering system offered by CLAAS. CLAAS Industrietechnik GmbH in Paderborn marketed the new technology after obtaining additional patents on switchgear and control valves to protect the complete system. Customers of the automated steering systems produced at this factory included other agricultural machinery companies who manufactured beet harvesters and harvesters for vineyards.

CLAAS maize headers today are fitted with automated steering systems which still deploy two sensor brackets to scan the crop rows. But there are some significant differences. Now outward facing, the sensors are capable of scanning virtually all row spacings typically found throughout the world, from 37.5 to 80 centimetres. Furthermore, a quick and intelligent signal processing unit greatly increases ground and harvesting speeds even on curves. And the system can even handle gaps in the crop. It recognises damage caused by wild boar, for example, or drilling misses, and still keeps the machine on track. So even today, the principle of the innovation introduced in 1975 is still minimising the operator's workload and maximising work rates.

SCAN + WATCH

"Our culture of innovation has remained unchanged since the founding of the company"

How has a successful climate of innovation arisen at CLAAS? How does one find the right employees? In search of the "CLAAS" code to unlock the secret of continuous renewal – a conversation with long-standing company helmsman Helmut Claas.

TEXT Edwin Baaske **PHOTOS** Andreas Fechner

› INTERNATIONALISATION
› CUSTOMER-FOCUS
› CULTURE OF INNOVATION

Time seems to have little meaning to Helmut Claas. Certainly there are no signs of the nine decades of this man's life as he talks about CLAAS today. When the 93-year-old speaks, there's a hint of boyishness to his manner. In the choice of his often pointed words. Or simply in the grin that plays around his mouth. Helmut Claas has mastered the art of sympathetic provocation. Time and again he tests the reactions, imagination and intellect of his opponent. The native Westphalian can put up with a lot. Except bores and people who tell him what they think he wants to hear.

Hemut Claas ran the agricultural machinery company for more than 50 years on the basis of this approach, and was for many years the managing partner. During that time he developed a culture of continuous progression, an innate commitment to improve – for the benefit of farmers and for the company. "Actually, this culture has remained unchanged since the founding of the company", says Helmut Claas, citing the founding members, the four Claas brothers, as the originators of the company's innovative spirit: "We have here a management style based on trust, loyalty, openness and" – the corners of his mouth lift slightly – "humour." You can immediately see that it's true. And he adds: "Innovative thinking has been the key to success at CLAAS right from the start."

He dismisses the idea that he set his sights well beyond the boundaries of the young Federal Republic at an early stage and created a culture of thinking ahead as a "pointless personality cult". If so, he maintains, it was largely due to the achievements of his father August and his brothers Franz, Theo and Bernard. "They drew on their tenacity, engineering skills and courage to build a company from nothing shortly after the First World War. "They were farmer's sons who had completed one year of higher education and gone on to gain an apprenticeship in metalworking and blacksmithing. They began manufacturing simple implements and machines, which resulted in early patents. The most successful invention was the development of a reliable knotter hook – the centrepiece of grain and straw binders. This generated the long-term capital the young company needed and was adopted as the company logo." Standing in the main building of this family-owned agricultural machinery manufacturer steeped in tradition, it's obvious how proud Hemut Claas is of his forefathers' achievements. In the early 1930s, the four brothers laid the foundations for what was to become a world-class company with the invention of the first combine harvester designed for European conditions. Today, we would call them first movers.

Helmut Claas has always regarded being in close contact and constant dialogue with employees as a key to success. Whether a hastily scribbled inspiration, concept ideas or approaches to marketing philosophies. Sharing ideas, in his view, is central to a vibrant innovation process. "When I took over the company, I learnt to see things in context", says HC, as he is affectionately referred to in-house. Over the years he has learnt that it's not enough to create innovations, you also have to communicate them appropriately. He believes it is important to have a clear company profile: to encourage, maintain and expand a sense of modernity, groundedness, sustainability and courage. "You have to look at things in context, look at the whole picture instead of the individual images. Everything has to fit within this context and be co-ordinated with each other", says the patriarch of agricultural machinery. "Today it is our image as innovators that differentiates us from the big American manufacturers." The unique selling point of CLAAS compared with the competition.

Now Helmut Claas is really warming to his subject. Strategic action is his field. "At CLAAS we have the advantage of being very flexible", he explains. "We can bring product innovations to market more easily and more rapidly than larger companies." CLAAS is starting from a smaller and more agile position

Expanding into North America: Helmut Claas during the mid-1960s when the company collaborated with Ford Tractors

"Our aim was and is to have a self-reliant workforce".

HELMUT CLAAS

Dialogue between the generations: Helmut Claas chatting to his daughter Cathrina Claas-Mühlhäuser.

than many competitors, he observes. This is an advantage, "especially as we know where our strengths lie". After all, it's easier to steer a boat than an ocean liner. Yet perhaps the most important factor in the company's success are the people who work there. "Our aim was and is to have a self-reliant workforce." New staff are never appointed purely on the basis of being particularly innovative or creative, he stresses. It is far more important "that they have ties with agriculture. The urge to innovate is all around us here in the air we breathe – we simply encourage that spirit of inventiveness." And how do you do that? "You have to set clear goals that encourage staff to think innovatively. They must know what their sector is all about and have plenty of room for manoeuvre." Helmut Claas has a good instinct. Yes, he would hold numerous individual discussions. Because these direct, personal conversations were motivating and gave him a sense of whether someone had the right level of commitment to the company. "It's important to identify with the company. A sense of ownership, of being part of our brand and our products. That's very pronounced in our company." It's also explains why so many family members of employees also work for the CLAAS Group – some in the third generation.

So clear goals, motivation and a strong sense of identification are all part of the innovative CLAAS gene. These characteristics have shaped the agricultural machinery manufacturer from the start. "CLAAS was already very large when my father died. All I did was continue to develop the company without thinking I was doing anything special. The product portfolio was also expanded during his era. He was responsible for introducing Seed Green for forage harvesters. The innovative first self-propelled forage harvester marked an important milestone in our forage harvesting machine programme." – "But", says Helmut Claas, "the original idea came from the employees", like many other innovations.

Is there a recipe for sensing when the time is ripe for an innovation? How do you know whether the market needs what you have invented? Helmut Claas pauses. For the first time this afternoon, he seems to be fighting an internal battle. Not for long, but for a perceptible moment at least, then his guiding principle of passing on knowledge gains the upper hand. But the answer is couched in the form of a gentle warning: "You have to be alert", slight pause, "and you have to be passionate about the subject, and enjoy taking a chance."

"The urge to innovate is all around us here – we simply encourage that spirit of inventiveness. For this you have to set clear goals. When everyone knows what their sector is all about and have plenty of room for manoeuvre, innovation happens".

HELMUT CLAAS

Paving the way to innovation

Without prototypes, there would be no innovative machines. Before full-scale production, early models show a development's true potential as well as harshly revealing those areas where further improvements are required – which ideas have a future, and which do not. The CLAAS family portrait gallery provides inspiring insights into the secret world of design in days gone by.

TEXT Alexander Bank, André Bosse **PHOTOS** Axel Struwe

CLAAS

ROLLANT RAPID: the world's first non-stop round baler remained in production for only a few years, but created the innovative net wrapping system which is now a world standard.

The idea of a powerful and efficient multipurpose tractor eventually evolved to become a CLAAS success story: the XERION.

› INNOVATIONS
› ONE OF A KIND
› PROTOTYPES

Successful innovations would be inconceivable without prototypes. They provide a playing field for engineers, a catalyst for a whole chain of innovations which make agricultural machinery substantially more efficient in subsequent years and thus embody a company's vision of the future. CLAAS history would be unthinkable without these sources of inspiration – the bedrock on which the company's success is built. The road from prototype to full-scale production has not always been short and direct, but test vehicles and machines have always shown that CLAAS thinks ahead before taking this important step.

The CLAAS tractor is a true milestone in the history of CLAAS prototypes: it was designed as a carrier vehicle for the LD-40 engine developed by Dr Fritz Schmidt. The engine's key data made it a giant among German agricultural machines in the late 1950s. At that time, no other German tractor manufacturer could boast an output of 72 hp. Dubbed the JUMBO, it proved its worth in-house initially and was gradually improved until the prospect of series production came closer to reality: Germany's agricultural industry demanded ever more powerful tractors, the well-known manufacturers didn't offer any machines beyond the 60-hp class and CLAAS products were highly regarded. The JUMBO had set the benchmark, but its scale-up ultimately failed because CLAAS simply lacked the necessary production capacity – such was the demand for harvesting machines at the time.

Sometimes global politics puts the brakes on innovation: The oil crisis and associated rapid rise in fuel costs ended the career of the APOLLO. The hot air dryer was the CLAAS engineers' brilliant answer to a pressing problem which threatened the very existence of many farmers: forage harvesting was only possible during sustained periods of dry weather.

In the 1980s CLAAS developed this small, versatile slope tractor for mountainous regions which was never mass-produced. However, elements of the FARMTRAC project have been incorporated into other developments such as the UNITRAC all-terrain forklift.

The HSG did not make it beyond prototype – and yet it is firmly established in the company's history: it was the bright spark that inspired the successful XERION family.

The APOLLO was a revolution in forage harvesting. The mobile green forage dryer produced energy-rich hay in briquette form – in the field, whatever the weather, in a fully automated process. Its fate was abruptly sealed by the oil crisis in 1973.

Prototypes are like a compass: they set the direction for subsequent series production.

The supercharged 1955 JUMBO powered by the CLAAS LD-40 engine was never in fact sold. Due to the strong demand for combine harvesters, there was simply not enough spare capacity to manufacture this tractor.

Too much moisture in the forage reduced its quality and in some circumstances even led to significant losses. CLAAS responded with a "giant hairdryer": In the early 70s the engineers developed the APOLLO dryer, which removed moisture from the harvested crop immediately after cutting and delivered an easy-to-handle forage briquette at the end of the treatment process. But the sudden rise in the price of oil made the diesel-driven APOLLO dryer uneconomical to use, so in the end, only 15 units were ever produced.

When it came to combine harvesters, CLAAS broke new ground with the JUMBO project (not to be confused with the 1950s tractor of the same name) and highlighted the company's innovative strength with the presentation of a convertible harvesting machine: with a fold-out grain tank, it only revealed it full size and harvesting capacity in the field. The combine harvester had plastic tanks to reduce the weight and a particularly low centre of gravity which allowed it to work on slopes, whilst movable functional elements facilitated maintenance and adjustment. The design of the JUMBO was also very striking: its smooth, rounded outer shell gave it a futuristic appearance which got the entire agricultural machinery sector fired up. However, series production ultimately failed due to the high cost of constructing the moulds. Virtually contemporary with the JUMBO, the DOMINATOR combine harvester series stepped in to take its place and was to subsequently prove a huge hit.

Crowd puller: CLAAS unveiled the JUMBO combine harvester at the 51st DLG Exhibition in Cologne in 1970.

Anyone nowadays scouring the market for small, versatile specialist tractors for mountainous regions will find vehicles inspired by a concept that CLAAS engineers were working on as early as the 80s. Back then they were developing the FARMTRAC slope tractor, a universal machine for forestry work and farming in mountainous and upland areas. Hydrostatic four-wheel-drive, suitable for gradients up to 70%, small turning radius, implement area with front and rear PTO, excellent stability thanks to a low centre of gravity and an extensive range of attachments: today's specialist tractors are constructed no differently from the FARMTRAC that was conceived a good 40 years ago. However, the versatile slope tractor fell victim to the company's success: CLAAS was operating at such full capacity that no scope was left to produce the FARMTRAC.

It's hard to find a better illustration of the increased efficiency achieved through innovation than the ROLLANT RAPID 56 – a round baler that could always beat the competition's time! While conventional machines had to stop in the field during the baling and bale ejection process, the tractor trailing the ROLLANT RAPID simply carried on. The electronically controlled feed unit developed by the CLAAS engineers allowed the crop to be picked up without having to stop. This meant that the baler's work rate was double that of rival products – a convincing argument for an agricultural industry at home and abroad that was operating under constantly growing cost and time pressures. The ROLLANT RAPID 56 launched in 1985 was well ahead of its time, but sadly this was also true of the installed electronic control components – during the mid-80s they were simply too expensive to make mass production a viable option.

A strong, sturdy carrier vehicle that provides a base for attachments for every season and every purpose: in pursuit of this ingenious idea, CLAAS developers had worked on the multipurpose 207 concept since the late 1970s. The 280-hp prototype 207 tractor developed by CLAAS served as the "drive platform". Equipped with quick release mounts, it was possible for two people to change the attachment in fifteen minutes without using tools. The basic vehicle fitted with special attachments would have worked out around 40% cheaper than the cost of purchasing self-propelled machines designed for seasonal use only and a separate tractor, and downtimes would have been correspondingly reduced to a minimum. The idea of versatility was certainly attractive and even though the 207 concept was not implemented initially, it signalled the way ahead for CLAAS: the idea of a powerful, versatile tractor evolved to become a CLAAS product which is a success story to this day: its name is XERION!

"There is still so much to discover"

The German Patent and Trade Mark Office (DPMA) in Munich is only a stone’s throw away from the famous Deutsches Museum. There, visitors marvel at yesterday's inventions large and small while at the nearby Office, more than 2600 employees are concerned with tomorrow's technical innovations. Lawyer Cornelia Rudloff-Schäffer, president of the federal authority for the past ten years, explains exactly what's involved.

TEXT PETER GAIDE **PHOTOS** LAURA THIESBRUMMEL

› INTELLECTUAL PROPERTY
› INVENTIONS
› PATENT PROTECTION

Mrs Rudloff-Schäffer, the German Patent and Trade Mark Office uses the advertising slogan "We protect innovations". How do you do that?

Well, firstly we rely on the applicants submitting their inventions to us. Over 85% of customers now do this electronically. Then we examine whether the invention really does go beyond the prior art because only then can a patent be granted. Patents must be novel, inventive and industrially applicable. Our examiners carry out extensive research, mainly in electronic databases, to determine novelty. An invention is deemed to be inventive if it is not obvious to a skilled person. As experts, our examiners are qualified to determine this autonomously.

Who are the examiners and what do they do?

By the end of this year we will employ more than 1000 patent examiners at the DPMA. They are all highly qualified experts, for example engineers, biologists and chemists. They have a very good understanding and a keen instinct for what is already out there, whether something will actually work and whether you could indeed manufacture it and use it on an industrial scale. And they have to assess all this when the invention is still at an early stage of development.

It sounds like painstaking detective work.

It often is. Each examiner handles between 300 and 400 open cases. They work with two computer screens, one showing the patent application and the other showing the global databases. China, Korea, Japan, the US – they have to search the 'prior art' everywhere. Using international patent classifications, we sift through around 100 million documents. Each examiner carries out research in their specialist field.

A patent procedure currently takes up to four years. Why does it take so long?

I am very keen for us to speed up the process. Patents protect innovations – but in order to be effective, the process obviously shouldn't take too long. We must strike a balance between diligence and speed. Our strategy is to complete the patenting process within an average of three years by 2030. To achieve this we need more staff, so last year the German Bundestag agreed to fund an additional 177 examiners, which is very good news.

Last year 36,000 patents were filed, only 43% of which resulted in a patent being granted. You seem to be very strict.

We are the largest national patent office in Europe and we have a very good reputation because our standards are high. Our examination process is prudent and accurate, but that is key to being able to assess whether an invention is really novel and worthy of protection. Our customers appreciate our thoroughness because ultimately, a company gains nothing from a patent that can be legally challenged.

To be clear, if I file a patent application at this office, the patent protection applies only to Germany?

Yes.

But your examiners still carry out research throughout the world. And if they discover something in Korea, for instance, the patent can no longer be granted in Germany?

Exactly, because it's no longer novel. The prior art must be reviewed. If you have a German patent, then you're already halfway there in many sectors. Not every company needs global protection. And nowadays technology is so complex that one company alone often cannot apply their intellectual property rights to every component. This is where cross-licensing agreements comes in. One company has a patent on a chip, the other on a different component and they agree to grant mutual rights to both parties' intellectual property.

So the notion of large, individual inventions taking over the world is somewhat outdated?

Well it's far less likely at least. The majority of inventions today are small further developments. Take IT or artificial intelligence, for example – collaborative partnerships and licenses frequently occur wherever software is involved. There are many things that an individual company can no longer develop in isolation. This has changed significantly in recent years – or rather decades. At the same time, we live in an era of disruptive developments associated with growing digitalisation which are bringing major changes to all sectors.

You have been very closely involved with patent law since 1984. What particular changes have taken place since then?

There is a continuing strong trend for patents to be filed by companies rather than individual inventors because research and development has become so complex and expensive. Only around 5.5% of the patent applications we receive are submitted by individual inventors. At the same time, patent applications worldwide continue to rise – 1.5 million a year in China alone at the last count.

And somehow you must take account of all these during the examination process?

That's right. The volume of prior art is growing exponentially. The flood of patent applications is really impressive. When I visited this patent office for the first time in 1991, one examination room contained individual drawers where hard copies of the publications to be searched were stored! That is inconceivable today. The databases that we now use were only in development back then.

China was for a long time known as the "workbench of the world". Will the country soon be at the forefront when it comes to innovations?

China has the political ambition to be the global leader in inventions. The increase in patent applications in China reflects an underlying strategy to promote innovation. Nevertheless, it is noticeable that the Chinese still file relatively few patents abroad. But the fact that you can file international applications to extend your reach beyond regional markets and into the world is one of the very things that make patents so valuable. German companies are still very strong in this respect. But the Chinese are catching up.

"Investment in research and development together with the number of patents filed worldwide are a good indication of innovative strength".

CORNELIA RUDLOFF-SCHÄFFER

Would you say that Germany can learn something from them?

There are more than 65,000 patents currently in force in Germany. That is a very significant number, especially when you consider the population size. That said, each country has its own tradition and its own approach. Germany's strengths lie in mechanical engineering, electrical engineering and chemistry. As Industry 4.0 connectivity becomes increasingly relevant, we must stay on task and play our part in shaping its future development. I'm very optimistic that we can do this.

To what extent is digitalisation already having an impact on the patent system?

Software is not patentable as such, only in conjunction with technology. This is a very topical issue. We are seeing annual percentage growth in double digits for applications in autonomous driving and artificial intelligence, for example. That's why we are taking on extra staff. Early in the year we set up new patent departments where we employ experts specifically to examine the link between software and technology. Our own work is also becoming increasingly digitalised. In the patent sector we switched exclusively to electronic files in 2011. In future, we will be using artificial intelligence in our research work.

Do you agree with the assertion that the more patents a company files, the more innovative it is?

I would say that investment in research and development together with the number of patents filed worldwide are a good indication of innovative strength. Only then does it become clear whether the patent will hold its value. Companies invest in other countries by appointing a patent agent in that country and shouldering the patent fees. That's expensive. Furthermore, companies can use patents to secure credit or loans. It is an asset.

Because successful patents affect the value of the company?

Yes, that's often the case. Take the pharmaceutical industry, for example. When a patent expires, it can affect the share price. It takes a great deal of time and money to develop blockbuster drugs. So to offset the expensive development costs, a relatively long patent protection is granted, which in some cases can even be extended. If this didn't happen, the company may have no incentive to develop new drugs.

What would the world be like if we didn't have patent protection?

Companies would probably try to protect everything using trade secrets. But how robust would that be? If you develop something nowadays, in principle you have to protect everything; the name – or brand – to distance yourself from others, the design, the wheels when you think about cars, smartphones and their touch functions, and utility models, which are a minor but nonetheless very important aspect of patent protection. Companies in many areas might even find it difficult to survive without the entire portfolio. I cannot imagine a functioning economy without protective rights.

Is it an exaggeration to say that patents even protect a society's ability to remain curious?

Curiosity is certainly something that lies deep within the human psyche and will always be there regardless of the framework conditions. But I think the patent system is still a very important driver. The reason that profit-orientated companies invest in progress and development is because it pays them to do so. They deliberately take calculated risks because they know that their successes will be protected. Nowadays more than ever, we need curious individuals who push the boundaries of our existing knowledge. Personally, I find it fascinating; inventors, scientists and academics tell us that there must be more, that we must make an effort and doggedly pursue our quest for knowledge. And they are right. There's so much more to discover out there.

FOUNDED IN BERLIN IN 1877 AS THE IMPERIAL PATENT OFFICE, THE GERMAN PATENT AND TRADE MARK OFFICE HAS BEEN UNDER THE LEADERSHIP OF CORNELIA RUDLOFF-SCHÄFFER SINCE 2009. THE LAWYER IS A WORLD-RENOWNED EXPERT IN THE PROTECTION OF INTELLECTUAL PROPERTY AND THE FIRST WOMAN TO HOLD THIS LEADING ROLE. THE 62-YEAR-OLD HAS BEEN ON THE BOARD OF TRUSTEES OF THE DEUTSCHER ZUKUNFTSPREIS AWARD SINCE 2009 AND CHAIR OF THE BOARD OF TRUSTEES OF THE MAX PLANCK INSTITUTE FOR INNOVATION AND COMPETITION SINCE 2015.

The best protection for good ideas

When you're constantly developing outstanding innovations, you have to expect copycats and imitators. Dr Steffen Budach, head of Patents and Trademarks at CLAAS, knows exactly how the company can protect itself – and explains in this interview how he and his team support the vital work of CLAAS inventors.

TEXT Marc-Stefan Andres **PHOTOS** Lukas Kawa

› PATENT STRATEGY
› INNOVATIONS
› INVENTIONS ACT

Dr Budach, what count as good ideas in your view?
We divide new ideas into two areas: inventions and innovations. Inventions are new technologies or processes which have yet to be applied. We call them innovations when they have evolved into something that can be sold. It's all about persuading the customer with a unique selling point.

What criteria do you apply to determine whether an invention is worth protecting?
We use a broad range of factors to assess the future viability of patents. For example, it's important that we use our research and development budget in the right areas, and this is also reflected in the patents. So we work closely with all areas of the company. At the same time, we analyse the profitability or market share of the invention in question.

Why don't you simply protect all inventions?
Well, it's a question of cost and efficiency. CLAAS has filed more than 4000 patents throughout its history. In 2018 alone there were 137 new registrations, which is around two-and-a-half each week. Many of these technologies are initially protected in a specific region only. Then after a year we consider where else in the world we may wish to file the patent. When working this out, we have to factor in the costs. A patent runs for up to 20 years, but we protect most products on average for twelve or thirteen years – after that, many technologies are no longer innovative enough to make it worth spending the money. We spend on average €25,000 to protect an innovation for the term of the patent. That's why monitoring the patents is another important aspect of our work. We perform a quarterly review to determine which are still current and which can be dropped. We have to use our budget as efficiently as possible.

Can you also protect business models which may be especially relevant to the latest developments in the field of digitalisation?
No, that's not possible because patents only apply to technologies. But we do identify business models that are important for the company and then run workshops to try to identify which features are critical to the process. We can protect these technologies and in this roundabout way protect the business model as a whole. Simply manufacturing good products is just as important. That's also a means of protecting an invention.

How does that work?
CLAAS has longstanding experience of working with customers and has established a relationship based on trust. They must be convinced that a business model developed by us, involving an app for example, will actually make their farm more profitable. So we must offer a guarantee of success. And then we don't necessarily need to protect this process.

Are the inventions which CLAAS protects always immediately transferable to products?
No, because many innovations take years before they are ready for series production. Experience has shown that completely new technologies need 20 years before they can be implemented. Nonetheless, we protect many ideas immediately because we take a long-term view. As part of our patent strategy we have reserved 30% of our expenditure for strategic topics, even though we don't know at the time whether they will work. Obviously we want to implement as many as possible. Our current implementation rate for strategic patents is 50%, which is good.

Can you illustrate this with some examples?
CLAAS has protected the continuously variable transmission for agricultural machinery – but the patent did not get off the ground initially. It was not until the basic patents had expired that we made sufficient progress to integrate the technology into series production. An example of a long-term patent is the APS HYBRID technology, in which we combined a tangential threshing system with a rotor. We were able to apply the patent immediately, which essentially gave us a 20-year monopoly on the market. The technology has now become standard.

Let's take a look at the latest product from CLAAS, the new LEXION: how many innovations are hidden away inside this combine harvester?

Dr Steffen Budach is head of the Patents and Trademark division at CLAAS. The qualified metalworker graduated in mechanical engineering before going on to qualify as a patent attorney. He joined Claas in 1997, and in 2001 took overall responsibility for patenting within the company.

We have filed 231 patent families for the LEXION. That means that each individual component features innovations, from the threshing unit, hybrid rotor and grain separator to the 3D cleaning system, sensors and multispectral camera.

And how do you make employees aware of the importance of patents?

Our team of patent attorneys, supported by administrative staff, travel around the world to engage with inventors and in the context of ideas management, encourage them to innovate. Engineers often believe that their innovation is too small to be protected. So we run patent workshops to convey the benefits of inventions. In many cases it's the small differences in the developments which offer something new compared with the competition. We always say that employees should share their ideas with us in any case – and we can decide together whether something is worthy of protection.

What motivates colleagues to come up with an invention and share their idea?

Many engineers have an inherent desire to be innovative. That's part of their job description, if you like. But it's also clear that inventors should benefit in some way from their invention, not just the company. At CLAAS we pay an inventor's bonus that operates throughout the world in accordance with the German Employee's Inventions Act. So all innovators receive a bonus.

Are there any significant differences between countries when it comes to patents?

On the whole, patent law is the same throughout the world. In China for example, it is very closely aligned with the German law. But we do have to examine very carefully which patents are of interest to which country or region.

Family as a driving force

Today, as yesterday, the success of CLAAS is dependent on the broad knowledge, gut instincts and technical expertise of its leading players. In this interview, Dr Patrick Claas, member of the Shareholders' Committee and the Supervisory Board, explains how the founders promoted a corporate culture based on trust, openness and innovative strength – and how the Claas family today continues to uphold these principles.

TEXT Edwin Baaske **PHOTOS** Thorsten Doerk

›TRUST
›TECHNICAL UNDERSTANDING
›INGENUITY

"My great-grandfather Franz Claas Senior and his four sons were genuine innovators", says Dr Patrick Claas, grandson of company founder Franz Claas Junior. "They shared a passion for new ideas and a fascination with technical solutions", explains the 55-year-old, who represents the interests of seven family members on his father Günther Claas's side within the company.

At the start of the 20th century the world became intoxicated with technology: there was a rush to invent machines which made people's lives easier, to recognise needs – and to turn these ideas into reality.

"My great-grandfather was an alternative animal practitioner", said Patrick Claas, who himself has a doctorate in physics. "So he visited farms every day, spoke to farmers and understood their concerns." The first machine that Franz Claas Senior designed was a centrifuge which made it easier to skim the cream from the milk. On the farm itself. "He did very well out of that", says Patrick Claas. This ability to trust gut instincts laid the foundations for everything that was to come later.

Essentially, the four founding brothers Bernard, August, Franz Junior and Theo were highly talented autodidacts. "If something worked, then it was improved. And if it didn't, they simply tried again." Machines were designed according to the principle of trial and error for several years. And then Fritz Schmidt from Berlin appeared on the scene: graduate engine design engineer and grandfather of Patrick Claas.

"He explained to me the mechanics of machines and such like", remembers grandson Patrick. One thing in particular stuck in his mind: "He would make drawings to explain exactly how something worked – and he could even deduce why it worked. Of course my grandfather also tinkered and experimented, but most of this work went on inside his head."

Fritz Schmidt impressed the Claas brothers with his theory-driven approach to developing new and often innovative products. In 1951 he applied to become chief design engineer for the newly established Engine Development division at CLAAS. Thanks to excellent references, he was soon appointed. And when Fritz' daughter Irmhild Schmidt married Günther Claas, the family ties were strengthened even further.

"My grandfather's approach was completely unknown in Harsewinkel at the time", says Patrick Claas. So the Claas brothers were both curious and impatient to see what Fritz Schmidt was developing in his drawing office. "They would often knock on his door because they were keen to see how his plans for the engine were advancing." And each time they were surprised to find that no test machine or experimental setup had been built. On these occasions, says Patrick Claas, his grandfather always gave the same answer: "I don't need to try anything out, I've done all the calculations and drawings."

For Patrick Claas, this example illustrates two things: "It's easier to put your trust in the unknown within the family unit, and openness to new, innovative ideas and approaches is an integral part of CLAAS DNA." The fact that the new invention was often also a success – as was the case with the engine – obviously played a part. "When my grandfather had completed the design, a couple of units were constructed from his drawings – and they worked first go." According to family lore, August Claas is reported to have said: "It's unbelievable, and all we used was paper." High praise indeed from the East Westphalian!

Trusting in the potential of employees is still a typical CLAAS trait. But to ensure that the right sort of products are designed and constructed in the first place, a second significant characteristic also comes into play: paying attention to customers and listening to what they have to say. "Alongside all the theory, which is essential if we are to continually offer cutting-edge technology in the 21st-century, at CLAAS we still remain particularly close to our customers." This proximity helps to distinguish the company from other global agricultural machinery manufacturers in the market. "We are an entrepreneurial family, real people made of flesh and blood. You can talk to us directly. At trade shows of course, but in-house too, you will receive a very warm welcome", says Patrick Claas. "We're not interested in short-term successes – we think sustainably, in the long-term." Continuity is one of the keys to success in the agricultural machinery industry, the other is innovation.

August 1954: Dr Fritz Schmidt (2nd from left) field testing the LD-40 engine he designed. Beside him on the right: Reinhold Claas.

> "We are an entrepreneurial family, real people made of flesh and blood. You can talk to us directly. At trade shows of course, but in-house too, you will receive a very warm welcome".
>
> DR PATRICK CLAAS

In future the aim will be to create stronger, better links between processes and to optimise entire process chains, whether it be technical processes in individual machines or between different machines or even management tasks. Connectivity, data management, machine automation and autonomous driving are some of the keywords in this context. Always with a view to safeguarding the profitability of agriculture from the vantage point of ecological sustainability.

One key aspect of this approach is that greater attention will be paid to the individual plant. Fields will be viewed more as compartmentalised, heterogeneous systems. The new technologies will make it possible to identify which parts of the field require more seeding or fertilising – and which less. This will be achieved with self-optimising machines that use sensors, laser technology, cameras and satellites to make processes measurable or to determine the condition and quality of plants – resulting in cost savings and more careful handling of soil as a resource.

1955, the LD-40 engine served as the drive for the new generation of CLAAS self-propelled large combine harvesters.

Digitalisation in farming creates an enormous volume of data. In future, it will be possible to analyse these data even more effectively using artificial intelligence and then convert them to recommendations or use them directly to initiate an automated optimisation process.

For example, thanks to modern camera technology it is already possible to reduce the proportion of broken grains in the harvested crop. This is done using algorithms which analyse images from a digital camera inside the machine. On the basis of these data, the machine settings can be optimised almost immediately.

Technology protects not only the soil as a resource, but also the human by reducing the burden of work. Whereas previously the driver would have had to monitor and perform several functions simultaneously under tough conditions, today modern technology provides a helping hand.

Agricultural engineering is part of Patrick Claas' life, and has been since his childhood. "Every evening I fell asleep to the sound of presses at work in the factory." Günther Claas and his family lived only a stone's throw from the factory, which the children used to visit on Sunday afternoons with their father

Born in 1963, Dr Patrick Claas studied physics in Bielefeld, Jena and Berlin. He represents the interests of Günther Claas' side of the family on the Shareholders' Committee and the Supervisory Board. His grandfather CLAAS design engineer Dr Fritz Schmidt had a formative influence on him. Patrick Claas has two daughters and lives with his wife in Gütersloh.

Together with his siblings, he runs his grandfather's farm near Osnabruck.

– the company was part of everyday life. "I grew up with five siblings so I know that family life is a delicate balance between harmony and conflict. A family consists of diverse relationships revolving around assertion, concession and compromise. But the older you become, the more you appreciate the common threads – in this sense, family is a blessing for personal development."

Differences of opinion can be found in any large family business, including CLAAS. But at shareholder meetings you get a clear sense from the start of what binds them. Everyone is aware of their responsibility for the whole organisation – and for the next generation.

"We all identify very strongly with the company." And the next generation is already being introduced to the CLAAS DNA. "In my generation, there was nothing of this calibre in place. Just the occasional crash course on shareholder meetings." It was fairly uninspiring – "apart from the times when we were allowed to drive the combine harvesters", says Patrick Claas with a grin. "But today it is far more hands-on and systematic, our children are perfectly prepared with specially tailored programmes and events. They have it a lot better than we did at their age."

INNOVATIONS
#16 – #30

If you take a close look at the culture of innovation that is such a feature of CLAAS, you soon discover what lies behind the development of successful innovations: innovators need inspirational ideas, perseverance – and the ability to keep sight of the benefits for customers at all times.

AXION
CLAAS
AXION

Rubber **instead of chains**

You need tractive power to harvest on difficult terrain. But how can you achieve that without compromising driver comfort or even being prohibited from driving on public roads? The answer lay in TERRA TRAC crawler tracks, which offered power and comfort while protecting the soil and increasing efficiency.

#16

NAME
TERRA TRAC
PRODUCT
COMBINE HARVESTERS
YEAR
1988

Crawler tracks and metal chains have been used on agricultural machines since the beginning of the 20th century. As farming became increasingly mechanised, farmers were beset by problems when harvesting on heavy soils in muddy conditions. In some areas tracked machines were the only option. CLAAS offered its first self-propelled combine harvester with steel tracks, but initially these were regarded as a niche product used mainly for rice harvesting. The metal chains were too expensive and too prone to wear to be more widely used. Furthermore, machines equipped with this drive technology were either prohibited from driving on public roads or very awkward to manoeuvre. With tyre manufacturers making major advances and continuously improving the performance of their products, agricultural markets were almost entirely dominated by wheeled vehicles.

SCAN + WATCH

But this was to change as another trend began to emerge. Agricultural machinery was getting bigger and heavier due to the introduction of new technology. Attention increasingly turned to the subject of ground pressure and soil compaction. At the same time, significant developments in rubber track technology were driving the transition from metal chains to rubber tracks and opening up a whole new range of potential applications. The reduction in ground pressure achieved by distributing the weight over a large contact area was the most important benefit for users. It minimised soil compaction and reduced yield losses. And unlike metal tracks, rubber tracks were also suitable for road use. CLAAS worked with Caterpillar initially on developments in this field. Ten years previously, the US company had developed a rubber track system for military deployment capable of transporting a 60-tonne load at up to 80 km/h. This expertise laid the foundations for developing the Challenger 65 crawler tractor. CLAAS and Caterpillar joined forces to develop a mobile track unit for self-propelled and trailed machines. The jointly developed Mobil Trac System (MTS) was a universal running gear that could be used on many self-propelled work machines. The first combine harvester to be equipped with this system was the COMMANDOR 116 CS, designed exclusively for the US market and very popular with American farmers.

The roadworthy full-track crawler, which used reinforced rubber tracks instead of the usual metal chains, could travel at a speed of up to 20 km/h without compromising the customary driver comfort since it was steered in the conventional way using the steering wheel. By the end of 1988, CLAAS and Caterpillar had introduced the name TERRA TRAC for the new profiled tracks. The US market responded even more enthusiastically to this innovative track technology when the Iowa State University published the results of a three-year study. This clearly showed that crawler tracks made a significant contribution to soil protection and thus ultimately led to improved profitability for the farmer.

Now it was time to capture the European market. Up until then, the new crawler systems had been designed as full tracks – but this fell foul of European traffic regulations, which specified a maximum width for vehicles on public roads. Furthermore, the electric steering systems used for the full-track crawlers were not approved for road use in European countries. To overcome this problem, the CLAAS developers began to explore a half-track crawler design.

Preliminary road and field trials in 1994 with a CLAAS MEGA 218 combine harvester had already shown that the half-track system offered significant benefits compared with the full-track version: narrow transport widths which would comply with European road traffic regulations, very smooth running characteristics and good pitch stability. In particular, approval for use on public roads was key to a successful European market launch.

In 1997, CLAAS launched the LEXION 450 with the first TERRA TRAC half-track series, which at that time was still rigidly attached to the axle. Each crawler track assembly consisted of four equal-sized wheels, two of which were rear drive wheels. The high-quality profiled continuous track belt was driven over the rear drive wheel by friction. The track belt was not designed to interlock with the drive wheel; lugs on the inside of the belt simply provided lateral guidance. In addition, a pair of suspended support rollers ensured better ground contour tracking and ground pressure distribution. Some of these features can still be found on the TERRA TRAC today.

Based on the success of this half-track system developed by CLAAS, in 1997 the company joined forces with Caterpillar once again to design modified half-tracks for the combine harvesters sold in the USA. These wider crawler track units were fitted with rubberised wheels and a new Caterpillar track belt. A slim-line version of this half-track crawler was launched on the European market in 2000. These new belts were far more hard-wearing and included an alignment device which kept them running smoothly. This innovation further increased the durability and service life. The installation of integrated axle suspension in TERRA TRAC crawler track units fitted to the LEXION 500 series in 2004 signalled a major innovation: the new suspension system effectively dampened any vibrations and jolts both in the field and on the road, and so prevented the machine from bouncing, just like wheeled machines. This solution delivered an even higher level of driving comfort by combining the technological benefits of both wheeled and tracked machines. Customers were now guaranteed soil-friendly harvesting and a more comfortable ride.

In 2006 the TERRA TRAC crawler tracks were reinforced with a one-piece cast frame to withstand the ever increasing weight of harvesting machines. In 2011 CLAAS engineers introduced a further innovative leap with the introduction of hydropneumatic suspension. Now all TERRA TRAC components had independent suspension. This crawler track system is still used today. The LEXION with the new TERRA TRAC crawler track proved a record breaking combination: it was the first combine harvester in the world to achieve speeds of up to 40 km/h on the road. Today around one third of all LEXION and half of LEXION HYBRID machines are fitted with TERRA TRAC.

The obvious advantages of the suspended TERRA TRAC track belts made the technology attractive for other applications too. The first machines to adopt the system appeared on the scene in 2017, when CLAAS unveiled two high-performance innovations at Agritechnica 2017 – the TERRA TRAC version of the JAGUAR 900 and the AXION 900. As the first half-track tractor with full suspension, the AXION 900 TERRA TRAC was awarded the silver medal by the DLG judges. The key innovation here was to adapt the suspended TERRA TRAC crawler track originally developed for the LEXION combine harvester to the requirements of tractors. The large drive wheel which allows high torque transmission for heavy traction work is instantly recognisable.

The prototype JAGUAR 900 with crawler tracks unveiled at Agritechnica 2017 was about one metre longer than the comparable wheeled version and had a modified front axle as well as a new gearbox. This extra length enables the machine to fully exploit the crawler track's many advantages. The JAGUAR TERRA TRAC is the first forage harvester to offer an off-the-shelf solution for protecting soil and grassland from machinery damage. The headline protection reduces the contact area by a third when turning and protects the grass sward. This makes it possible for the first time to use the advantages of a crawler track system on all surfaces the whole year round.

Having proved its worth in combine harvesters since the 1980s, the CLAAS TERRA TRAC concept has now been transferred to tractors and forage harvesters.

Crop flow
Unbeatable efficiency

The success of the JAGUAR forage harvester is underpinned by a conveyor system patented in 1980 designed to improve crop flow. By hugely reducing the amount of energy required and increasing the efficiency of the crop flow system, this new technology proved a game changer in forage harvester design.

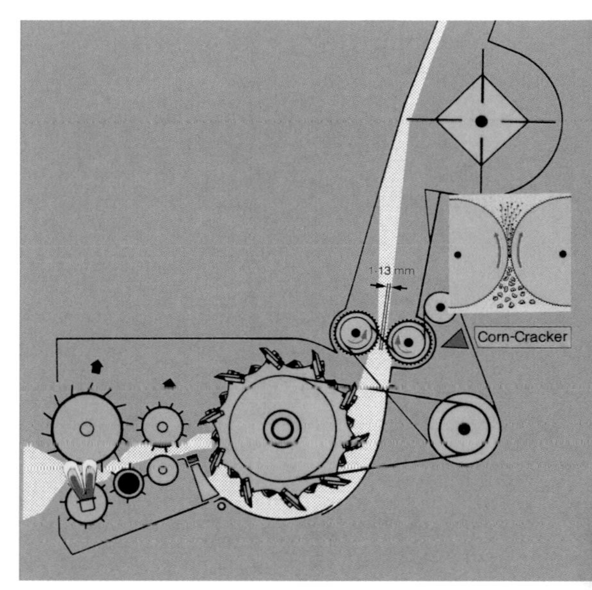

The crop flows in a straight line through the machine without slowing down.

#17

NAME
CROP FLOW
PRODUCT
FORAGE HARVESTERS
YEAR
1983

In self-propelled forage harvesters the crop is drawn into the machine from the front attachment, pre-compressed and then conveyed through the first of two openings into a rapidly rotating knife drum. Here the crop is chopped into small pieces and then discharged at high speed through the second opening into the discharge chute and from there into the forage trailer via the discharge spout.

This simple operating principle was adequate for small machines because the distance from the discharge spout to the forage trailer was relatively short. But the bigger forage harvesters became, the more exacting the discharge performance had to be. One option to improve performance was to install a blower between the knife drum and the discharge chute. But this approach wasted a great deal of energy. The blower consumed a significant proportion of the engine output in reliably conveying the chopped material to the transport trailer due to friction and other inhibiting effects.

The quest for greater efficiency has always inspired CLAAS developers to come up with new ideas, and here it was no different. They abandoned the power-hungry blower that was commonly used as a means of conveying the chopped material to the trailer. And came up with an exceptionally energy efficient alternative – the crop accelerator: a paddle wheel mounted in the discharge chute with four hardwearing paddles. As a result, the energy required to discharge the crop remained low even with high throughputs. Instead of having to negotiate awkward angles as with conventional systems, with this design the crop flowed through the machine in a straight line with increasing momentum. This brought about a significant improvement in performance compared with blower-type machines of equal power as well as substantially reducing energy consumption for comparable performance. The innovative process set new standards in forage harvesting technology when it was introduced in the JAGUAR 600 series in 1983.

SCAN + WATCH

AUTO FILL
Right on target

The CLAAS JAGUAR is a powerful forage harvester which demands a great deal of concentration from the driver to prevent crop losses. The AUTO FILL system is the ideal assistant. It safely takes care of unloading, and also makes it possible to work round the clock.

AUTO FILL uses digital 3D image analysis to enable automated discharging from the forage harvester to a trailer running alongside.

#18

NAME
AUTO FILL
PRODUCT
FORAGE HARVESTERS
YEAR
2009

During harvesting operations, the CLAAS JAGUAR 980 forage harvester can produce more than 300 tonnes of forage per hour. This equates to more than one eighty kilos of chopped material discharged through the spout every second. Previously, it was the driver's job to drive the JAGUAR and, at the same time, control the spout to ensure that the chopped material landed safely in a trailer running alongside or trailed behind the forage harvester. If one of the drivers momentarily lost concentration, hundreds of kilos of forage could be inadvertently dumped in the field instead of in the trailer in a matter of seconds. The CLAAS engineers saw it as their job to develop a technology that prevented this scenario. The result of their endeavours was the AUTO FILL automatic trailer filling system, which was awarded a DLG gold medal in 2009.

The automatic control of the discharge spout to fill the trailer is based on 3D optical pattern recognition and image analysis. A high-resolution camera mounted on the discharge spout does not simply capture the trailer driving alongside; it measures it and detects the fill level. The system continually analyses these data, processes them in real time and generates control commands to direct the discharge spout. Not only does it recognise specific trailer types and their fill levels, it can also direct the spout to any type of transport trailer. Furthermore, since AUTO FILL can direct the spout lengthways and sideways, the driver can determine where the crop will land in the trailer in advance.

While initially the AUTO FILL function only enabled discharging from the forage harvester to a trailer running alongside, a further development made it possible to discharge automatically to transport vehicles driving behind the JAGUAR. The camera image with overlaid, real-time filling level indication is displayed on a monitor in the JAGUAR cab for the driver, who can intervene at any time. Depending on the crop for example, he may decide not to completely fill the trailer. Swivelling work lights on the discharge spout allow the camera to carry on working even in the dark – a feature that makes 24-hour harvesting possible.

SCAN + WATCH

The ORBIS folds to a width of 3.0 meters for road journeys and thanks to the integrated transport system, the driver can travel safely on the road in full compliance with road traffic regulations.

ORBIS
Ingeniously folds away

The row-independent ORBIS maize header paired with the JAGUAR make the perfect team. In addition to the operational reliability of the technology, the ORBIS offers customers convenience throughout the working day: it takes just 15 seconds to switch to transport mode for fast, safe road journeys.

#19

NAME
ORBIS
PRODUCT
FORAGE HARVESTERS
YEAR
2018

Contractors are always looking for machines which satisfy not only their customers' high quality standards, but their own demands for speed and efficiency as well. The ORBIS meets these criteria: the front attachment for forage harvesters that is mainly used for maize was introduced by CLAAS over ten years ago. In 2018 the developers unveiled a completely re-engineered generation.

As well as a host of minor improvements, the ORBIS now features a newly designed folding mechanism. Now available in four versions with working widths ranging from 4.50 to 7.50 m, the maize header can now be folded quickly and completely symmetrically, making it possible to drive the forage harvester on public roads with ease.

The front attachment is designed mainly for use in maize but can also be used to harvest other crops such as miscanthus or sorghum. The largest version weighs an eye-popping 3400 kg. This weight indicates just how robustly engineered the harvest attachment is – from the frame to the contoured, press-hardened construction of the T-panels. This greatly reduces the cutting angle in relation to the ground to ensure shorter and more uniform stubble heights.

In view of its weight, it's impressive how quickly the ORBIS maize header can switch from working to transport mode: the mechanical and hydraulic systems require just 15 seconds to fold a 7.5-metre working width to a roadworthy width of just 3.0 metres. Watching the folding process from the front, you get the impression that you're looking at a bird elegantly folding its wings as it lands on the ground.

With the maize header in transport mode, the driver can work just as comfortably on the road as in the field. The new ORBIS enables fast, safe on-road travel at up to 40 km an hour, and with the maize header folded away, the driver has very good visibility. The enhanced comfort and convenience of the JAGUAR/ORBIS combination ultimately reduces transfer times between fields, giving the contractor more time to focus on the harvesting process.

CLAAS ORBIS maize headers are designed and produced at the Bad Saulgau production plant, which specialises in forage harvesting machines. In one of the largest grassland areas in Europe, some 500 experts develop, test and produce mowers, tedders, swathers and forage trailers. Bad Saulgau also produces all the front attachments and crop flow components for the JAGUAR.

SCAN + WATCH

JAGUAR
CLAAS

Running like **a well-oiled machine**

No forage harvesting fleet would be complete without a swather – an implement which gathers the mown crop into rows known as swaths. The CLAAS LINER rotary swather family, which deploys one or several rotors, has proved particularly popular. A system patented in 1993 that keeps the machine running smoothly and protects it from all kinds of soiling marked a major advance in swathing technology.

The multiple gearing attachment keeps the tine arms firmly seated, while the patented PROFIX bracket mounting makes them easy to remove and replace.

#20

NAME
ROTOR DOME ASSEMBLY
PRODUCT
FORAGE HARVESTING MACHINERY
YEAR
1993

The individual rotors of a rotary swather are like a merry-go-round in which the seats move up and down and tip back and forth as well as going round and round. This interaction creates a wave-like motion which can be seen in each of the 14 tine arms of the swathers in the CLAAS LINER family. The physical forces exerted on each arm by this motion are considerable. And what's more, they vary not only as the tines rotate but also depending on the harvested crop, soil conditions and other parameters. Dusty or damp conditions, soiling and flying forage debris are a further complication. All these factors have one thing in common: they increase wear on the swather.

A fully encapsualted rotor dome assembly was the solution: all the tine arms terminated in this assembly, which was driven by an extremely robust cam track made of spheroidal graphite cast iron capable of withstanding enormous forces. The rotor dome and cam track were hermetically sealed and filled with low-viscosity grease. This arrangement ensured that the cam rollers and all moving parts ran smoothly with minimum wear. CLAAS first fitted this innovative solution to a new CLAAS LINER series in 1993. The benefits of the new technology soon became apparent and the market success proved that the developers have made the right decision. The performance of the LINER 770 was particularly impressive: the test machine swathed a total of 8000 hectares in two harvests, yet despite this enormous volume of work needed nothing more than a tyre change. Apart from that, the mechanics remained fully intact and no further maintenance work was required. This clearly demonstrates the strength of the technology. It kept on working round-the-clock without a hitch.

Since 2008, the company has deployed an enhanced version of the CLAAS rotor dome assembly: the developers changed the lubricant inside the transmission from low-viscosity grease to oil and improved the connection and durability of the tine arms. The aim behind every further enhancement of the rotor dome assembly is to prevent wear, enhance comfort and reliably protect the core components from soiling.

SCAN + WATCH

Overview
Wheat
normal
unknown
inactive
CEMOS
DIALOG
AUTOMATIC
CRUISE PILOT
Help
Settings
35 %
dry
14 mm
8 mm
610 rpm
16 mm
1020 rpm
Cleaning
t/h
125
89
%
CEBIS
17:16
89 %
82 °C
92 %
1020 min-1
670 min-1
5.8
km/h
65
30
23.40 ha
3.58 ha/h
9.76 t/ha
35.00 t/h
STOP

It's all about the **settings**

"You know it's an innovation when the market shouts hurray!" is a long held marketing maxim. But what if the innovation is so far ahead of its time that customers initially respond with scepticism rather than enthusiasm? This is what happened to the CLAAS engineers when they first launched CEMOS AUTOMATIC. But then the tide turned, and the rest is history, as they say.

#21

NAME
CEMOS AUTOMATIC
PRODUCT
COMBINE HARVESTERS
YEAR
2012

The "CLAAS Electronic Machine Optimisation System" – CEMOS for short – unveiled at Agritechnica 2009 is a dialogue-based system that guides the operator to the best combine harvester settings for the current harvesting conditions and then continually adjusts them. The assistance system runs on an external terminal which helps the operator select the best settings for the job in hand in three steps using a screen dialogue: first the operator requests a suggestion for a setting (for example to reduce grain losses). Then CEMOS suggests specific settings. The operator can accept or reject these suggestions and repeat this dialogue until he is satisfied with the results. The new settings – a different fan speed, for example – are then performed by CEMOS, provided that the change is confirmed by the operator. No adjustments are made automatically without operator confirmation.

CEMOS DIALOG also allows the operator to request suggestions for mechanical adjustments, using tools if necessary. A harvesting machine can have up to 50 setting options, depending on its level of equipment. Only the most important ones can be adjusted from the cab. The operator may have to stop the machine to adjust the cutter bar if crop flow problems arise during difficult conditions. This is an example of a mechanical adjustment.

SCAN + WATCH

CEMOS works with preset values derived from years of practical experience. These CLAAS data are based on proven average values for virtually all harvesting conditions. However, since these values are usually a compromise, there is often plenty of scope for optimisation. Thanks to CEMOS DIALOG, the operator can exploit this potential to the full.

A record-breaking performance produced by a LEXION 770 TERRA TRAC on 1 September 2011 demonstrates the outstanding results that can be achieved from LEXION combine harvesters equipped even with the first-generation CEMOS. The operators set a new world record that was officially confirmed by Guinness World Records representatives at the time: they harvested 675.84 tonnnes (22.5 percent higher than the previous record) in just eight hours. That was equivalent to a throughput of 84.48 tonnes per hour with a fuel consumption of 1.15 litres per tonne (down 10.8 percent). The machine went on to harvest a total of 1361.99 tonnes during some 20 hours of continuous operation. With a fuel consumption of 1.20 litres per tonne, this corresponded to an hourly throughput of 68,1 tonnes. Despite this record-breaking achievement, the CLAAS developers were not prepared to rest on their laurels. Next came the CEMOS AUTOMATIC. This innovation was the logical development of the CEMOS dialogue-based system but with a completely new software concept which was simple to use yet highly effective. Once activated, the software sets the relevant rotor speed and rotor flap position for residual grain separation in the combine harvester automatically. In the cleaning system, CEMOS AUTOMATIC controls the fan speed and the upper and lower sieve openings and continuously adjusts them to changing conditions. A LEXION equipped with this technology together with GPS PILOT for automatic steering and CRUISE PILOT for ground speed and throughput cotrol came very close to the vision of automated harvesting. One of the strengths of CEMOS AUTOMATIC is its ability to continuously re-adjust the settings on a scale that no operator would be able to achieve manually. The operator chooses a harvesting strategy that can be changed at any time, and the machine takes care of the rest. So the combine harvester performance is optimised at all times, while the operator keeps a watchful eye on the whole process – no more and no less. And because the integrated digital functions ensure that the machine settings are optimally configured, the harvester achieves the maximum possible harvest yields for the chosen strategy. There are four different optimisation strategies to choose from: maximum throughput, minimum fuel consumption, high threshing quality – or a balanced ratio.

If rain is approaching, the operator can instruct the machine to harvest as quickly as possible, accepting moderate losses in the process. If the weather stays dry, CEMOS AUTOMATIC adjusts the harvesting machine accordingly by applying a particularly fuel-efficient strategy or focusing on maximum threshing quality.

The operator can of course choose to disable CEMOS AUTOMATIC fully or partially at any time. This step may be useful when harvesting special crops, for example, or in extreme harvesting situations. He then has the option of continually adjusting the residual grain separation and cleaning system manually while CEMOS AUTOMATIC is temporarily disabled. Pressing the autopilot button on the multifunction lever is all it takes to reactivate CEMOS AUTOMATIC so that it takes control again.

When using CEMOS AUTOMATIC in the field, farmers soon want to know what settings are required to avoid broken grain. Or to put it another way: does the system adjust particularly sensitive settings automatically? At first, the threshing drum speed and concave clearance remained a matter for the boss. The operator decided on a strategy and CEMOS DIALOG suggested improvements. Then the dialogue-based system became more sophisticated – it suggested a modified setting, the driver approved it and then checked the results. A quick glance to the grain tank at the rear was enough to assess the proportion of broken grain, for example. If he was satisfied, he could retain the new setting. If not, he requested CEMOS DIALOG to make a new suggestion or alternatively, he made a manual adjustment.

This principle changed in 2017 when CLAAS launched the new CEMOS AUTO THRESHING. As a fully automatic control system for the drum speed and concave clearance of a tangential threshing system, CEMOS AUTO THRESHING marked a significant step towards automated combine harvesting. The system starts with a balanced basic setting, but then, using state-of-the-art sensors, it adjusts the concave clearance and threshing drum speed dynamically to suit the changing threshing conditions. So this new system ensures that the threshing unit is always used to its full potential, the desired grain quality is achieved and no time-consuming presetting is needed.

By automating the final combine harvesting assembly in the machine, CLAAS has transformed machine operation: moving away from adjusting the settings of individual units towards specifying an overall agrotechnical strategy. Nowadays the operator no longer has to grapple with the complex effects of optional settings on the harvest result. Instead, he specifies his top priority: grain quality, threshing performance, cleanliness or chop quality. The harvesting machine derives the maximum harvesting performance on the basis of this specification.

That also sounds great, but how do experienced operators respond to it? Do they perhaps get the impression that a digital technology is trying to get the better of them? The truth lies in the field: Comparative test drives were undertaken during the first harvesting season that CEMOS AUTOMATIC was available (which predates the introduction of CEMOS AUTO THRESHING in 2017). Drivers with decades of harvesting experience operated their CLAAS LEXION either with or without CEMOS AUTOMATIC – when they stepped down from their cabs, it was immediately clear from their mood, the grin on their face, their look of amazement or a slight gnashing of teeth that CEMOS AUTOMATIC worked! The system made their job easier, improved efficiency and also offered inexperienced operators the chance to push a high-performance combine harvester in the CLAAS LEXION family right to its limits without risk. So it turns out to be true – you know it's an innovation when the market shouts hurray!

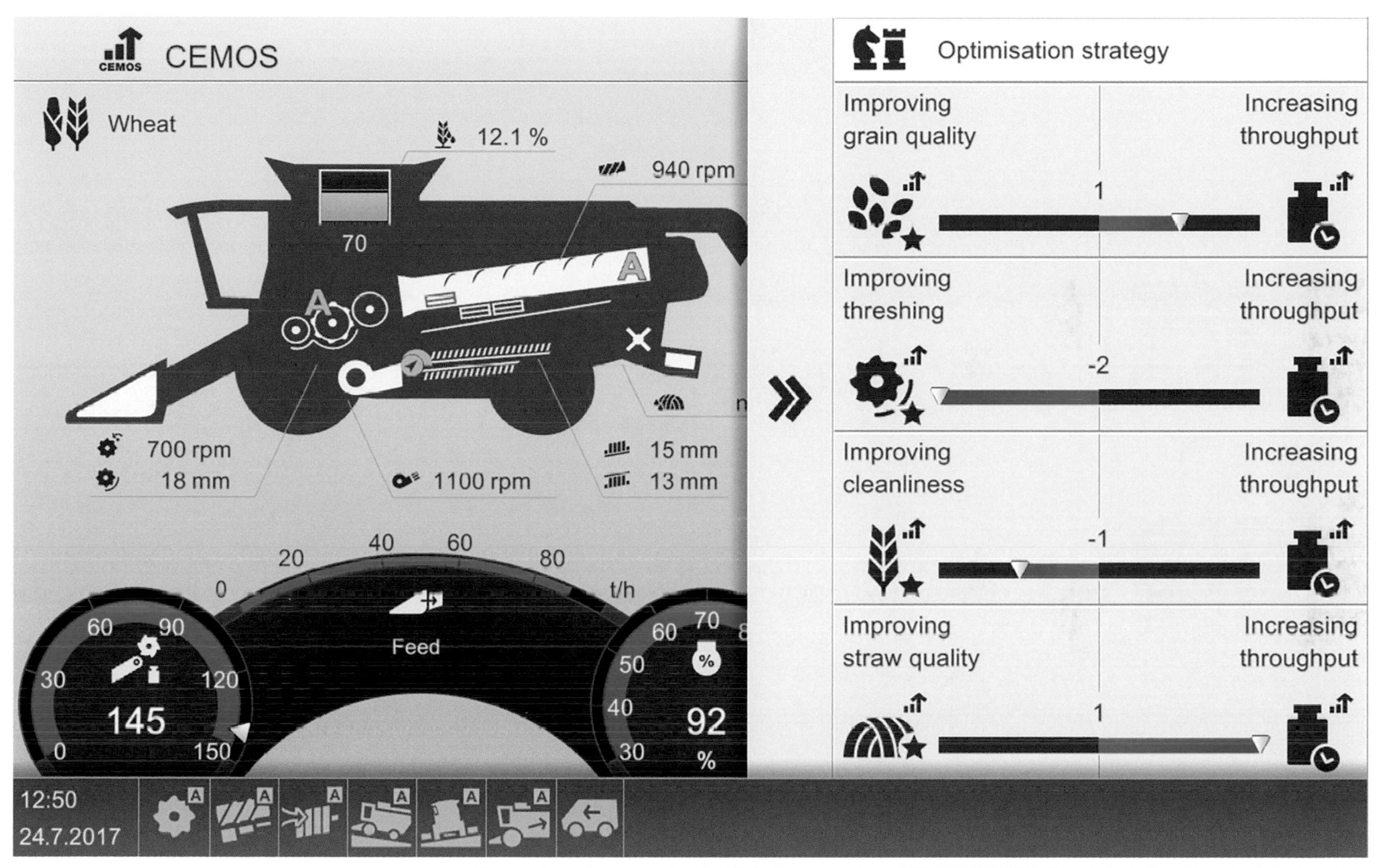

The operator makes a selection which indicates the results that CEMOS AUTOMATIC must aim for to achieve the desired strategy.

Nowadays the operator no longer has to grapple with the complex effects of optional settings on the harvest result. Instead, he specifies his top priority.

CLAAS
VOLTO 1050

Tedder **with patented folding mechanism**

The rotary tedder is a forage harvesting implement that plays a big part in producing high quality forage. The rotating tines on the machine spread the freshly mown grass over the field or turn the harvested crop. The aim is to ensure that the crop is evenly dried before it is swathed and brought in.

So big, and yet so small. The VOLTO 1050 folds up very small for road transport – a system patented by CLAAS.

#22

NAME
VOLTO 1050
PRODUCT
FORAGE HARVESTING MACHINERY
YEAR
1997

Early tedders were developed in the 19th century, but it was the 1960s before the first rotary tedders appeared on the scene. Driven by the tractor's power take-off shaft, these implements soon became highly effective harvesting assistants. With their help, a job that already had to be done in a short timeframe was completed faster. But the implements were generally wide, too wide in fact, and that created a problem: German road traffic regulations specified a maximum width of 3.0 meters and the same was true of most European countries.
Some manufacturers tackled the problem by offering folding tedders with more than six rotors. However, a great deal of time and effort was required to operate the folding mechanisms available at the time. The development of the CLAAS VOLTO 1050 at the CLAAS production plant in Saulgau solved the problem: the machine had eight rotors and a working width of 10 meters. For road transport the tedder retracted to a compact 2.97 meters – in compliance with the width limit. This was achieved with a patented folding mechanism which enabled the tedder to automatically retract from working position to transport position using multiple folding movements. Once in transport position, the tedder had a comparatively low centre of gravity, allowing it to travel on roads at up to 40 km/h. The wheels used for road transport were fitted with conventional car tyres.

In addition to this patented folding mechanism, the CLAAS VOLTO 1050 also offered a night swathing function, which was particularly useful for haymaking given the tight timeframes. Once swathed, hay absorbs much less moisture overnight and so dries much faster. Other manufacturers only offered this function with a special gearbox which was costly and time-consuming to install. With the VOLTO 1050 the driver simply had to adjust the PTO speed on the tractor.
The CLAAS VOLTO 1050 marked a milestone in the development of rotary tedders. Inspired by its success, the developers at the Bad Saulgau plant came up with several further innovations. So today's tedders have ten rotors, a 13-metre working width and the MAX SPREAD crop flow concept – the industry standard which is now fitted to all models in the VOLTO range.

SCAN + WATCH

10.91 ha/h
0.25 ha
1002
90 bar
10.6m
110 bar
110 bar
1002
1002
V00.99.00
CLAAS
CEBIS
C3

One app **for everything**

In the first phase of digital farming, machines were equipped with their own terminals and all data were pooled in the user's office. The UT app from CLAAS allowed farmers for the first time to use their own tablet as a universal device.

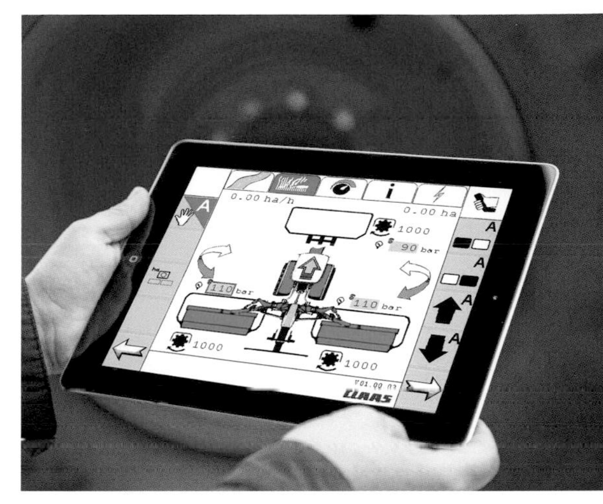

The UT app became available from 2016 as the EASY on board App. It allowed all ISOBUS-compatible machines to be conveniently controlled from a tablet.

#23

NAME
UNIVERSAL TERMINAL APP
PRODUCT
TRACTORS
YEAR
2013

In 2013 CLAAS received a gold medal at the SIMA Innovation Awards for its Universal Terminal App (UT app). This fact in itself shows just how much the industry has changed – that one agricultural machinery manufacturer received an award for developing an app highlights the importance of digitalisation to CLAAS and how successful the company has proved in this field.

The key to developing operating devices for harvesting machines and tractors has always been to design the interface between human and machine so that the operator can get the best performance from the machine. Terminals installed in the machines structured this interface using ISOBUS, a standard communication protocol for the agricultural industry. The terminals allowed communication about working processes and, depending on the equipment installed, data transfer between machines and the farm or contractor's computer, which served as the central desktop computer. The operator had to make changes to the machine's system settings via the on-board terminal display. The CLAAS UT app revolutionised the human-machine interface. With the new app, for the first time users were able to use their existing tablets as machine terminals. It was now possible to display and use the ISOBUS system on a conventional tablet. This level of compatibility offered users a whole host of new opportunities – in particular they now had access to all existing operating data, applications and other apps at any time. Contractors, for example, could access customer files, emails or weather data whenever they chose.

This development signalled a paradigm shift: with this CLAAS app it was no longer necessary to equip each individual machine with a terminal. The focus changed from hardware to software. Nowadays operators have one universal tablet which they can use for different machines and on which they can view all the data. All from the comfort of their cab.

SCAN + WATCH

The innovative bolted construction guarantees maximum deflection and impact resistance while avoiding the weakening effect of welded connections.

MAX CUT
Simply a better cut

The MAX CUT mower bar is fitted on all CLAAS front mowers and almost all DISCO rear mowers. Because this technical development has to be one-hundred-percent reliable, the company tested it worldwide working closely with customers. The feedback was very encouraging and a successful market launch soon followed.

#24

NAME
MAX CUT
PRODUCT
FORAGE HARVESTING MACHINERY
YEAR
2015

CLAAS DISCO mowers can either be used singly or as a triple combination with one front-mounted mower and two rear-mounted mowers. The name DISCO is a reference to the cutting discs mounted on the bar which runs along the ground. Disc mowers are mainly used for forage grass, but another common application is the production of biomass for anaerobic digesters. The core component of the mower is the mower bar. Its design influences not just the resulting forage quality but also the economic efficiency of harvesting operations. The critical factors here are the volume of forage harvested in proportion to the time expended, how operator-friendly and wear-resistant the machine is and how quickly the knives can be changed.

In 2015 CLAAS introduced a new generation of mower bars under the name MAX CUT. Depending on the model, individual MAX CUT mower bars are supplied in widths from 2.60 and 3.80 meters with six to nine mowing discs. The bed of the mower bar is made from pressed steel plate cut from a single piece which is bolted instead of welded to give the overall structure high torsional strength and stability. The mower bed features a wave-shaped leading edge combined with an ingenious system of finely ground gear wheels to ensure efficient power transmission. This wave design also enables the mowing discs to be placed well to the front to achieve optimum crop flow.

These technical optimisations together with specially shaped connecting pieces increase the mower's cutting surface – a design that guarantees optimum cutting results in all operating conditions. Furthermore, MAX CUT mower bars can be run at 850 rpm. The rotation acts like a built-in economy power take-off, reducing diesel consumption by up to 15 percent.

Like many other CLAAS developments, MAX CUT has a very direct impact on the customers' job. So it was especially important for the developers to conduct pilot tests in the field before planning series production. Prior to introducing MAX CUT, around 1000 of these innovative mower bars were field-tested worldwide. The test farms included contractors in New Zealand, Italy and France as well as farmers in Germany, France and Japan. Feedback from these customers and other testers provided a pool of knowledge which the developers used to make further improvements and modifications. Thanks to these tests, the product that was eventually marketed met the needs of customers worldwide – launched as an innovation in DISCO mowers, it proved a hit right from the start. The judging panel for the Steel Innovation Prize were equally impressed. In 2018 they awarded MAX CUT second place in the "steel products" category.

SCAN + WATCH

A **comfortable ride**

When CLAAS acquired a majority stake in the agricultural machinery division of the French company Renault in 2003 and went on to fully integrate tractor production within the CLAAS Group five years later, the company also acquired new colleagues who brought with them a wealth of expertise. This included a patented four-point suspension system for the driver's cab – an innovation that CLAAS was to develop further.

The comfort afforded by the "Hydrostable" four-point cab suspension introduced by Renault in 1987 was legendary. CLAAS perfected the technology.

#25

NAME
HYDROSTABLE CAB SUSPENSION
PRODUCT
TRACTORS
YEAR
1987

French cars in the 1960s and 70s were fabled for their comfortable suspension. As anyone who has ever driven a Citroen DS (pronounced "Déesse", meaning godesse) can surely testify. And no doubt this unique selling point of French cars is what inspired two engineers from Renault Agriculture to register their invention with the European Patent Office in 1987: a four-point suspension system for tractor cabs.

The underlying idea was to decouple the cab from the tractor's rigid chassis and suspend it from four damping elements. This special suspension design went far beyond a well-upholstered and adjustable seat – it effectively absorbed vertical and horizontal vibrations at four points. As a competitive advantage, this innovation was worth its weight in gold and when Renault Agriculture was renamed CLAAS Tractor following the takeover, the new parent company adopted the full suspension system, making it an exclusive feature of its range.

The innovation was certainly good for the driver's back, but its true value went well beyond that. If good cab suspension means that you still feel fresh after a hard day's work, you will make fewer mistakes – and if you make fewer mistakes, you achieve better results. The logic of this argument was a key selling point, so it made sense to further improve the four-point suspension after acquiring Renault Agriculture. Indeed, it was a matter of urgency, since competitors were working on very similar suspension systems.

The CLAAS refinements involved the mechanical design of the four-point suspension: initially the four suspension points were identical but the developers soon changed this, separating the two front suspension points from the rear ones to increase scope for adjustment. This development produced a further, significant increase in driver comfort in what was after all their workplace. The other change affected the entire tractor suspension: the suspended and electrohydraulically controlled front axle constructed in the Le Mans factory and deployed in tractors over 90 hp played an important role in this development.

Worthwhile improvements were also made to the front and rear linkage suspension, but the most important advance was to neatly synchronise all the separate suspension elements – including the driver's seat, which featured low-frequency suspension. This complete package offered drivers unrivalled comfort, a more efficient, safer working environment and protection for the machine. The innovation was also sustainable: even today, the four-point cab suspension system is still the benchmark for comfortable suspension in tractors.

SCAN + WATCH

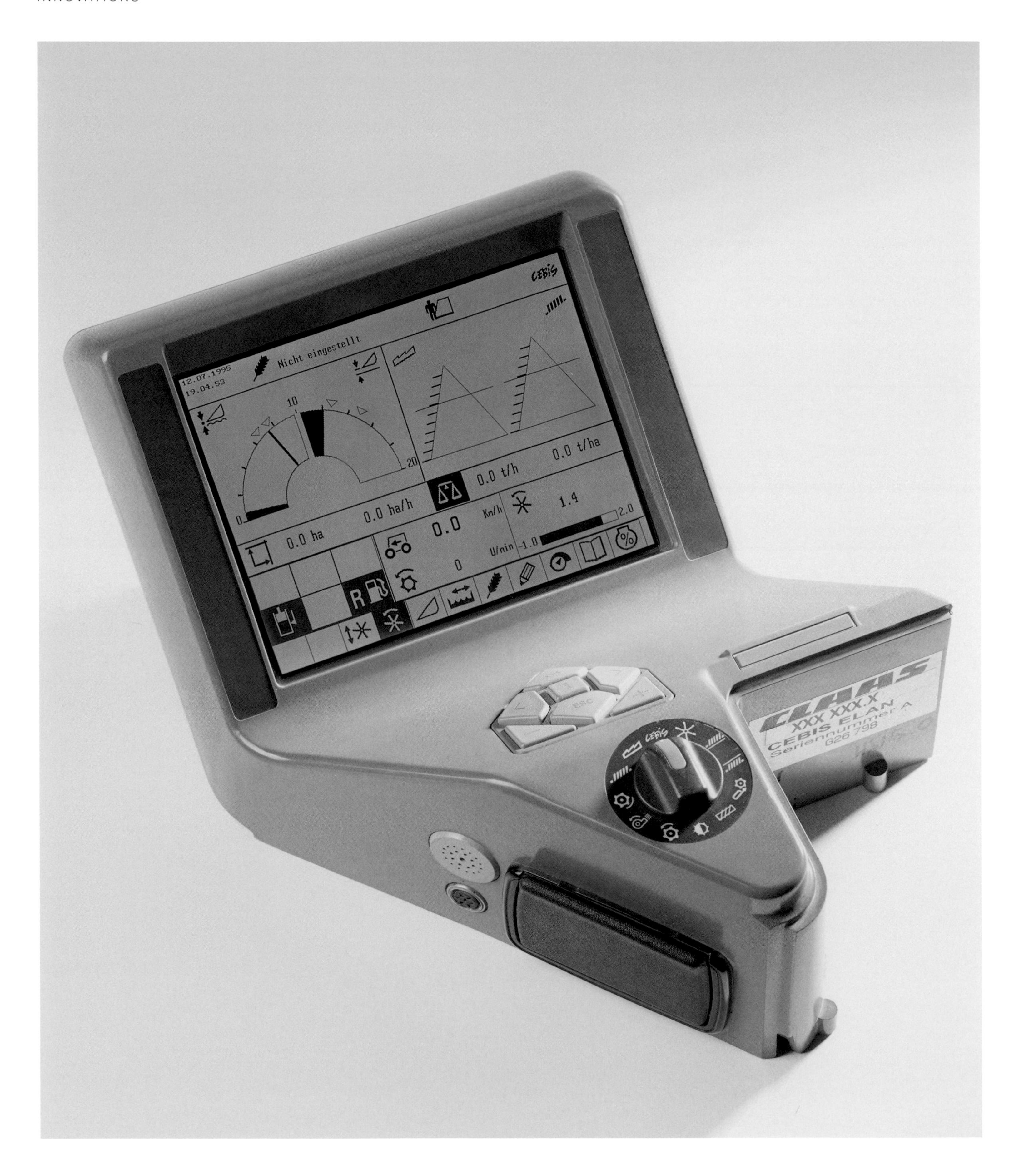
CEBIS
Nicht eingestellt
12.07.1995
19.04.53
10
20
0
0.0 t/h
0.0 t/ha
0.0 ha/h
0.0 ha
0.0
Km/h
1.4
2.0
U/min
-1.0
0
R
ESC
CLAAS
XXX XXX.X
CEBIS ELAN
Seriennummer A
026 798

CEBIS
The brain behind the network

Combine harvesters are often described as "factories on wheels". Complex processes take place inside these machines which instead of being housed in an air-conditioned factory are moving across the fields. So it's crucial to provide the operator with optimally processed information about every conceivable parameter. This is exactly what the CLAAS electronic on-board information system CEBIS does.

CLAAS set new standards in human-to-machine communication when it introduced the CEBIS on-board information system.

#26

NAME
CEBIS
PRODUCT
COMBINE HARVESTERS
YEAR
1995

Rotational speeds, sieve settings, grain separation quality, ground speeds, throughputs and moisture contents – the list of parameters that are relevant to operating a combine harvester is long. CEBIS ensures that all measurable parameters are presented in an appropriate form, and can be controlled and where necessary recorded.

The CLAAS fieldwork computer was originally entrusted with this task. With the more sophisticated CEBIS, the aim was to process all the available data in such a way that it could be viewed and if necessary, directly controlled from a single terminal. This required a standard data bus system to allow devices to communicate with each other – the CAN bus developed by Bosch in the 1980s in collaboration with chip manufacturer Intel. CAN stands for Controller Area Network, subsequently standardised internationally as ISO 11898. CLAAS equipped its LEXION combine harvesters with this system to provide them with an on-board data network.

Imagine CEBIS as the brain behind the network. It obtains a wide variety of information via the data bus system which in the first-generation devices was displayed on a ten inch black-and-white LCD display. As well as this display, the first CEBIS complete system included an industrial PC, a PCMCIA card reader, a small printer, an external keyboard and a buzzer which signalled alarms and notifications. Even at this early stage of development, the list of functions covered 15 areas, including the direct manual setting of all individual units in the combine harvester, yield mapping, job monitoring and cutter-bar height control.

But the first-generation CEBIS was just the start of fast-paced development in this field. CLAAS has now pooled its expertise in the interface between electronics, digitalisation and mechanical engineering in the subsidiary CLAAS E-Systems based in Dissen. In the development centre opened in 2017, software and hardware developers, design teams and agricultural engineers work on innovative solutions, including continuously improving the range of options available with CEBIS and its ease of operation.

SCAN + WATCH

The CMOTION multifunction control lever: the hand and arm rest comfortably in an ergonomic position on the armrest.

Everything **in hand**

The developers at CLAAS began to turn their thoughts to machine control concepts at a very early stage. Their developments in this field culminated in the ergonomic CMOTION multifunction control lever unveiled in 2011. But the road was a long one.

#27

NAME
CMOTION
PRODUCT
TRACTORS
YEAR
2011

Even the first self-propelled HERCULES combine harvester introduced in 1953 had a three-way hydraulic valve that could be shifted via a three-phase shift gate. This convenient arrangement meant that the driver could use the same lever to control the operating speed, header and reel. But to do this he had to use the strength of his upper body, right arm and right hand to push the lever from left to right or back and forth.

Later, as CLAAS machines became increasingly powerful, operating them became somewhat more challenging because the driver had so many more units to control and shift. As hydraulic and electronic systems increasingly replaced mechanical ones, the force that had to be applied to each unit diminished, but the fact that the levers and switches were not always located close to the driver's seat called for a different form of physical exertion.

A multifunction control lever for combine harvesters introduced by CLAAS in 1985 was an important milestone on the road to achieving today's levels of comfort. The company fitted this control element – which resembled the joystick of a helicopter – to several combine harvesters and later to the forage harvester range as well. Over time, further functions from the right-hand console were migrated to this lever. But it was not until the launch of the CMOTION in 2011 that the goal of concentrating as many elements as possible within the lever to make life easier for the driver was finally achieved.

During the development phase it became clear that too many control buttons can overwhelm the user and also result in "mouse arm" or repetitive strain injuries. With this in mind, CMOTION is designed so that the arm and hand rest in a comfortable, ergonomic position on the armrest or lever. The driver can control all the important functions using his thumb, index and middle finger.

Now the developers had another problem to deal with: since the driver's hand hardly ever let go of the lever during the working day, in direct sunlight it began to sweat. They resolved this issue by incorporating a ventilated grille into the surface of the lever.

SCAN + WATCH

F7
F9
F10
F8
ISB
AUTO

CLAAS
JAGUAR

The **mighty** JAGUAR

CLAAS has built more than 40,000 JAGUAR forage harvesters since 1973. The name of the series was not chosen at random. Jaguars have the strongest bite of all the big cats. In 1994 the developers made a giant leap forwards with the introduction of the JAGUAR 800 series, combining performance with reliability, comfort and versatility.

#28

NAME
JAGUAR 800
PRODUCT
FORAGE HARVESTERS
YEAR
1994

The rapid rise in maize cultivation in the late 1960s created a demand for increasingly powerful harvesting equipment. The trailed forage harvesters available at the time were unable to provide the required levels of productivity, especially as field sizes increased. Some contractors and small engineering firms began to put together the first self-propelled forage harvesters using existing assemblies. Among these was a CLAAS supplier who named their machine the "Imperator". For CLAAS, the logical progression was to put these developments on a professional footing and so the company set about constructing the first self-propelled forage harvester in Harsewinkel. Since CLAAS was able to draw on existing trailed JAGUAR models and proven combine harvester components, the project quickly came to fruition.

The first JAGUAR: double-row header, 120 hp

In 1973 the company presented the results: the JAGUAR 60 SF, a forage harvester with a double-row maize header, 120 hp, driven by a rear-mounted engine. By 1976, CLAAS had manufactured more than 500 of these self-propelled forage harvesters. Its successor, the JAGUAR 70 SF, was launched for the 1976/1977 season with a choice of two engine sizes, 150 and 175 hp. In 1975, between the introduction of these small forage harvesters, CLAAS launched their big brother – the large 213-hp JAGUAR 80 SF. This machine had a significantly wider chopping cylinder plus the ability to separate the feeder and chopping cylinder housing for easy servicing, a feature which is still present in the JAGUAR today. The list of equipment also included a new blower, while the CLAAS automatic steering system and a heated, ventilated cab were available for the first time as optional extras. For CLAAS, driver comfort was no less important than machine performance. The introduction of the considerably larger JAGUAR 600 series in 1983 signalled a further major investment by CLAAS in technical innovations. Even then, self-propelled forage harvesters were deployed in maize around 70 percent of the time. Equipped with a new six-row maize header which folded inwards to comply with the permitted width for on-road use, a new corn cracker above the chopping cylinder which cracked the maize grains between two rollers, a new energy-efficient accelerator in the discharge chute and many other features, the series was a winner from the start. Unit sales of almost 7000 machines made CLAAS the world market leader with a market share of more than 50 percent. Five years later in 1988, CLAAS introduced a completely redesigned JAGUAR 600 series. The top-of-the-range 354-hp JAGUAR 695 SL established a new performance class, while the V chopping cylinder with offset, half-knives ensured even crop flow and reduced clamping forces in the accelerator.

SCAN + WATCH

800 series: a quantum leap

Following a development period of more than five years during which time the engineers experienced a series of highs and lows, 1994 saw the launch of the JAGUAR generation that was to be a benchmark for many years to come and even today still sets the standards for forage harvesting technology: the 800 series. Alongside standard harvesting headers, this featured the first row-independent maize headers and a cutterbar for mowing and chopping whole-crop silage in a single pass. The JAGUAR 800 was not only more powerful than its predecessors, but now, for the first time, it was possible to attach eight-row maize headers. The reliability and maintenance friendliness of the machine had also been significantly enhanced. The engine, for example, was installed crosswise so that it could be easily accessed from all sides. This design gave the driver an advantage: in preparation for harvesting grass silage, for instance, he could simply replace the corn cracker with a grass shaft in a matter of minutes. This operator-friendly design feature was a trailblazing development for modern forage harvesters worldwide. Customers also appreciated the central lubrication system because it cut down on maintenance and so saved them a great deal of time. Power was transmitted from the engine's crankshaft directly to the chopping unit, corn cracker and crop flow accelerator via a special main drive belt called a powerband. The long-lasting, maintenance-free main clutch was an additional advantage: the innovative design of the direct drive proved so durable that it was subsequently transferred to the JAGUAR 900 series without any modifications, accompanied by the marketing slogan "We haven't changed a thing!".

The engines of the 800 series achieved maximum output at 1800 rpm, so fuel consumption was correspondingly low. The top-of-the-range JAGUAR 900 delivered an output of 605 hp – and this power had to be cooled. A large radiator screen with rotating dust extraction system provided the necessary cooling. Intake air was drawn in from above from the direction of travel, through the radiator and then routed over the engine and discharged from an air outlet at the rear. In previous models it was possible for the hot air to enter the cab, turning it into a sauna on hot days. This was now a thing of the past.

Comfort in the cab

The JAGUAR 800 series were and still are aimed mainly at contractors. Recognising that even the very best forage harvester is only as good as its operator, the developers made cab comfort and ease of operation a top priority. CLAAS had the advantage of being able to adapt the recently developed new combine harvester cabs for the forage harvesters. These new cabs were equipped with a second seat: a handy feature for contractors' clients, who instead of watching from the field margin could now ride up in the cab with the operator. And the fact that the operator could control many settings from the comfort of the cab saved time and made forage harvesting operations more efficient.

More power and speed on request

Certain models in the JAGUAR series offered customers additional exceptional optimum values. The legendary JAGUAR GREEN EYE launched in 2006 delivered an astounding 623 hp along with numerous innovations to boost efficiency and user convenience. In 2003 the JAGUAR SPEEDSTAR was added to the diverse JAGUAR 800 series. With a top road speed of 40 km/h, this fast version halved the travel time from field to field, enabling it to match the pace of the tractor-trailer combinations that made up the rest harvesting fleet. In February 2019 the 40,000th JAGUAR rolled off the production line in Harsewinkel, but the JAGUAR TT didn't do this entirely on wheels – this most recent model features the latest generation crawler tracks. This further innovation is a symbol of the fact that the JAGUAR uses cutting-edge technologies to deliver on its commitment to performance and reliability.

"It was only later that we recognised the value of the product name JAGUAR. It became almost a byword for high-performance forage harvesters".

HELMUT CLAAS

World leader. Over 40,000 JAGUAR forage harvesters have been manufactured since 1973.

CLAAS
CLAAS
CLAAS XERION 3300
OPTITRAC
GOODYEAR

The versatile **powerhouse**

Whether an innovation works or not depends on a whole host of circumstances. The history of the first CLAAS large tractor series illustrates the importance of taking the long view when it comes to good ideas – and biding your time until the time is right.

The overall concept of the XERION, including its versatility, makes it unique in the industry.

#29

NAME
XERION
PRODUCT
TRACTORS
YEAR
1997

HUCKEPACK was the product name and concept for a multifunctional vehicle developed by CLAAS in the 1950s, which consisted of an implement carrier with a threshing unit mounted on top. After the grain harvest, the farmer could separate the two components and convert it to an all-purpose tractor-like vehicle. With its large choice of attachments for transport, tillage and other activities, the HUCKEPACK could be used all through the year. However, with technical developments in the industry – not least from CLAAS itself – focusing increasingly on specialised high-performance machines, the HUCKEPACK went out of production in 1960. But the company never lost sight of its vision of a self-propelled tractor as a versatile platform.

Towards the end of the 1960s, CLAAS engineers launched the HSG project initially as a means of developing a hydrostatic transmission for combine harvesters ("Hydrostatisches Getriebe" in German, hence the abbreviation HSG). But HSG soon morphed into a project in its own right, providing CLAAS with an entry ticket into the large tractor segment. Prototypes were developed but after a few years CLAAS put the HSG project on ice: with an economic crisis calling for action to be taken, HSG fell victim to a savings programme. This decision was prompted partly by the fact that a planned collaboration with Daimler-Benz had recently come unstuck. The idea had been to jointly develop a large tractor, the industrial and municipal version of which would be distributed by Daimler-Benz and the agricultural version by CLAAS. The car manufacturer was to launch the MB Trac one year later, with no involvement from CLAAS. But despite these setbacks, CLAAS saw no reason to abandon the idea of having its own large tractor.

In 1978 work on the forerunner to the XERION began under the name Project 207. Many earlier ideas – especially that of an all purpose vehicle as a platform for different applications – were now revisited, redeveloped and implemented using technically advanced solutions. The XERION was born. Now deployed throughout the world, this large tractor is not limited to forestry and agricultural work, but has long been used in airports, opencast mines and other industrial applications too. Today the XERION is not only a flagship model but also a symbol of the fact that really good ideas don't simply get forgotten when external developments initially stand in their way.

SCAN + WATCH

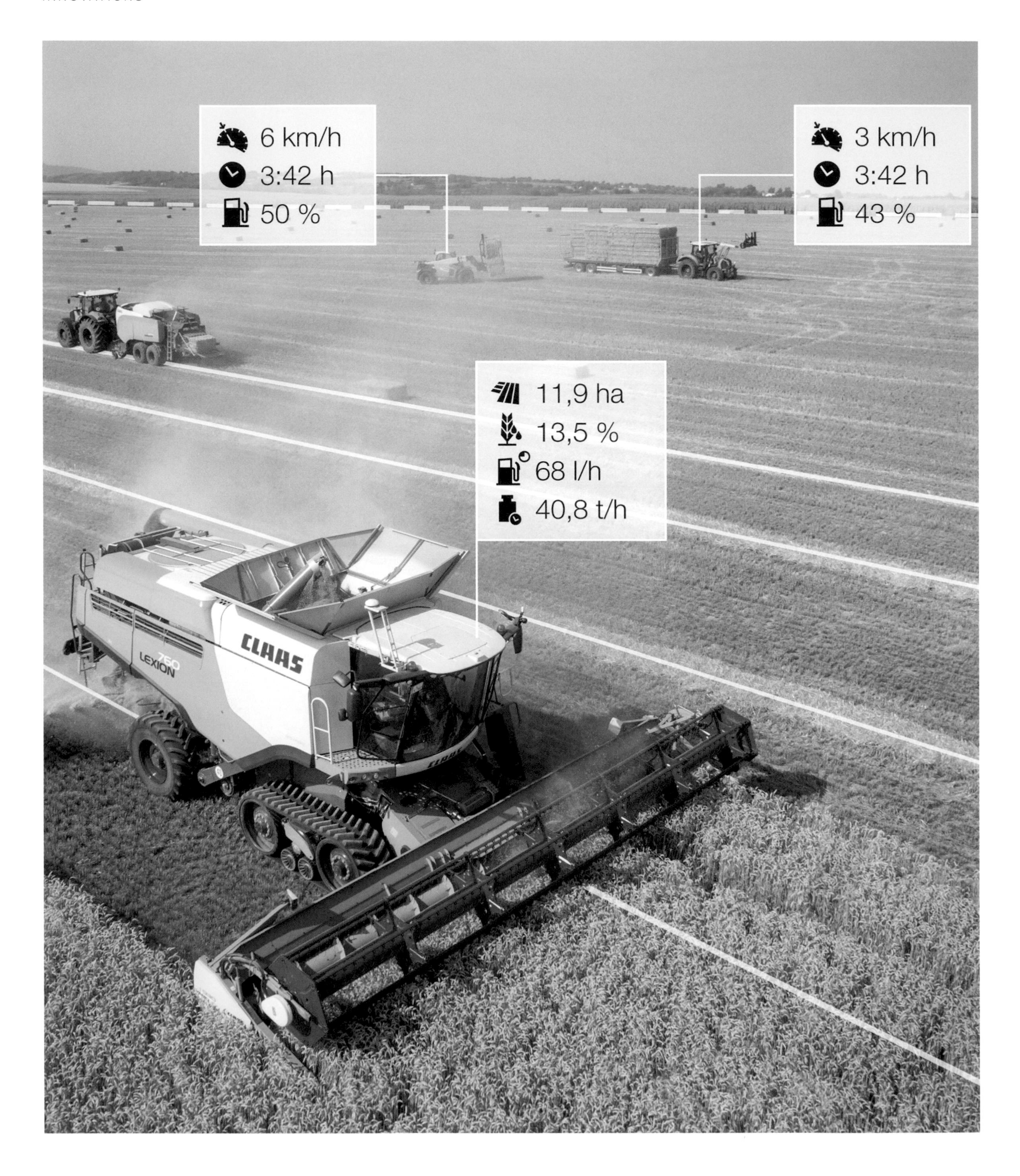
6 km/h
3:42 h
50 %
3 km/h
3:42 h
43 %
11,9 ha
13,5 %
68 l/h
40,8 t/h
CLAAS
LEXION
750

And then came TONI – **the CLAAS TELEMATICS evolution**

Telemetry, the automated transmission of measurements generally over large distances, was first used in the 1920s for weather balloons. Manufacturers also used early telemetry systems in agriculture, although for testing and development purposes only. All that was to change in 2005. With digitalisation creating new opportunities, CLAAS launched its own, highly advanced telemetry system which customers could use themselves: TELEMATICS.

Getting the whole picture: all key machine data can be quickly accessed on the farm PC.

#30

NAME
TELEMATICS
PRODUCT
COMBINE HARVESTERS
YEAR
2005

The terms "precision farming", "smart farming" and "Agriculture 4.0" encompass a range of systems designed to provide professional data management. Nowadays, farmers need access to accurate data recorded and transmitted in real time to deliver professional farm management and telemetry is key. Machines equipped with this system transmit over 200 different parameters by mobile phone to a central computer. The producer always retains ownership of the data. Via a highly secure server, these data can be instantly accessed on the customer's computer, which is usually in the office of the farmer, contractor or farm manager. So from the office, farmers can not only monitor and analyse workflows and processes on the farm, but optimise many processes as well. For example, they can pass on recommendations for machine settings during ongoing harvesting operations. The ability to analyse data going back several years is particularly useful because it provides a more secure basis for planning and sequencing future projects.

CLAAS has achieved a great deal since introducing TELEMATICS for the 2005/2006 season. LEXION combine harvesters were the first machines to be equipped with TELEMATICS. Then CLAAS gradually added TELEMATICS to the AXION 900 and 800 tractors, the large 4x4 XERION tractor, the JAGUAR forage harvester and the TUCANO combine harvester series. In addition, in the European Union, a factory-fitted SIM card was incorporated into these machines to transmit data by mobile phone.

And then came TONI. The name stands for "Telematics on implements" and describes the connection of attached implements, e.g. balers or forage trailers, to the tractor and to the TELEMATICS system using the ISOBUS tractor and implement communication interface. CLAAS developed TONI in conjunction with other agricultural machinery manufacturers. Since the 2012 season, TONI has included a feature which is particularly useful for contractors. They can print off the machine's performance data in the form of a delivery note using the on-board printer and submit it directly to the customer on completion of the job. These data are transferred to their own accounting system at the same time.

At Agritechnica 2013 CLAAS launched a completely redesigned TELEMATICS website and the new TELEMATICS app for use on mobile devices such as smart phones or tablets. Ten months later, CLAAS released three different versions of its TELEMATICS system. So now there was a system perfectly tailored to the needs of each individual customer. And true to the CLAAS doctrine of never being satisfied with their achievements, the development of CLAAS TELEMATICS is still ongoing.

SCAN + WATCH

"Innovation? It's built into our very walls!"

As chair of the Supervisory Board at CLAAS, Cathrina Claas-Mühlhäuser always has a view of the bigger picture. In times of digitalisation, disruption and skills shortages, it's a huge challenge that the granddaughter of company founder August Claas relishes. Because she knows that she can rely on the innovative strength of the employees, their solutions-oriented approach and the well-functioning corporate culture.

TEXT Marc-Stefan Andres, Thomas Lötz **PHOTOS** Andreas Fechner

› CORPORATE CULTURE
› PEOPLE
› TECHNOLOGY

Cathrina Claas-Mühlhäuser appreciates clear statements. In her view, the role of chair of the Supervisory Board at CLAAS goes beyond her committee work. She sees herself as custodian of the corporate culture, which for her also underpins the innovative CLAAS spirit. "When we celebrated our 100th anniversary, we asked ourselves what the critical success factors were which had enabled us to remain in existence for so long. Time and again, we came back to the corporate culture", explains Cathrina Claas-Mühlhäuser. "We have a particular way of doing things at CLAAS which sets us apart: people learn from other people here, and so their knowledge and values are passed on from generation to generation." This spirit, the inquisitiveness of employees and not least, the lively culture of debate must be protected. And engaging with one another on an emotional level – purely for its own sake – creates a deep and unique sense of solidarity. "If we retain our solutions-oriented approach, we will also retain our inventive spirit. My father has always said 'It's built into our very walls.' And it's true."

With her father Helmut Claas at the helm during the second half of the previous century, the company evolved to become the world's leading agricultural machinery manufacturer. Cathrina Claas-Mühlhäuser, born in 1975, completed an apprenticeship in commercial administration with another company and then studied business administration at the University of St Gallen in Switzerland. After a trainee programme, she worked for the tech company ABB in Zürich. Then at the age of 29 she returned home to join the Supervisory Board at CLAAS. She was appointed deputy chair of the Shareholders' Committee and has chaired the Supervisory Board since 2010.

"I received an open and friendly welcome when I joined the company", she recalls. "The CLAAS team spirit helped to ensure that I acquired the necessary knowledge over the years to be able to take up this position." Her colleagues helped her along the way, she relates: "It was important that I too was open, it's the only way to develop good new ideas together." Innovation is always a team achievement, even today when the challenges are very different: "My father always says that in the past, everyone seemed to be able to turn their hand to anything. Today we are reliant on more specialist knowledge than one individual could possibly acquire." And, most importantly: innovation occurs on every level within the company and in every sector, not only in relation to the product. "Processes are almost as important now, especially when it comes to making the journey to the destination as short as possible or non-core processes such as planning, accounts and HR as efficient and cost-effective as possible", says Cathrina Claas-Mühlhäuser. The job of management is to create freedom and space which allows creativity to flourish in the first place. Otherwise it's easy to get bogged down in bureaucracy, especially when the business is growing.

As far as Cathrina Claas-Mühlhäuser is concerned, corporate management involves the management board and the supervisory board working together. This does not necessarily imply major changes: "If you go about it the right way, many employees only notice the changes gradually and can focus on their tasks. There may be more radical reforms, but these should be handled in such a way that after a year they are simply taken for granted." Change in any event also requires trust and a certain level of continuity as regards the general direction. A system must also be in place to facilitate communication. "If our only concern is who came up with the good idea first, people will quickly stop talking to one another. The only important thing is that it is implemented!"

One of these ideas is the pioneering farm management system 365FarmNet, which Cathrina Claas-Mühlhäuser has supported from the start. "Similar software packages were already available, but based on the outdated Windows system and only for the German market – the team working on this project saw the development of an integrated system as a major advance. And only the management could release the resources to do this." So innovation passes from hand to hand – as long as the channels of communication run smoothly.

At CLAAS, innovation is not simply ordered from on high by decree. "Instead it comes about when people find a different, faster, better and more efficient way of doing what they do." However, this requires the best qualified, most talented people – which in Germany at least is gradually becoming more difficult. In China, Russia or the US, young people are more mobile and highly motivated. In the CLAAS subsidiary in the Russian town of Krasnodar for example, welders have relocated from Vladivostok 3000 kilometres away to work there. It's not just

Tradition and innovation: trust, knowledge and experience provide the basis for new ideas.

because of the business model that 365FarmNet is based in Berlin rather than the company headquarters in Harsewinkel, but also because this cosmopolitan city is home to far more digital specialists.

Is the new world order causing jobs to be lost? "Absolutely not!", says Cathrina Claas-Mühlhäuser vehemently shaking her head. "Although most 'new' jobs do require a sound education. By that I don't just mean a degree – often, an apprenticeship is even better because it is practice-based. But we have to offer something to make us appeal to the new generation and to set the right priorities." Young people's expectations of work are different today than for their parents' generation. They want flexibility in terms of location and hours: one only wants to work four days a week, another is happy to work five – as long as three of them are from home. "It's difficult for an industrial company like ours to reconcile these different working time models," says Cathrina Claas-Mühlhäuser. "Some activities here can only be done in situ – you can't take the assembly line home with you!" With other jobs however, a more flexible approach can and must be encouraged, she adds. "Unfortunately it's not always possible for people to take this on board because not everyone wants to be 'totally flexible'; they need routine. And that's absolutely fine too", says Cathrina Claas-Mühlhäuser. "As long as they don't prevent the others from being more flexible." And clear job assignments are essential, as development processes in particular can no longer be constrained by a simple framework. This increases the demands placed on all managerial staff. "But that's how we create space for innovation."

The truth of this assertion is backed up not least by the emergence of the Greenhouse: a co-working space that has recently been created in Harsewinkel. The start-up furnishings were not ordered from some hip company in Berlin, but mainly produced by the trainees and staff at CLAAS. With the help of colleagues from Production – and mostly after clocking off from work. And the setting up of the Greenhouse has initiated more creative ideas: from now on every apprentice, every intern and every trainee will do a coding course. "Because being able to use an app is not the same as being totally digital", says Cathrina Claas-Mühlhäuser. "The point is: I must be in control of the machine. Not the machine in control of me." Only then can digital innovation succeed. "If I know how it works, I can make things happen and make processes more efficient. It won't be easy getting the entire workforce to understand this – but with the Greenhouse, we have made a good start."

"Tuesday Unlimited" has also been launched as a contemporary take on further education: on the first Tuesday of every month, between 100 and 200 people from all parts of the company assemble in the Greenhouse. "Then someone gives a talk about 3D printing, virtual reality or artificial intelligence – all it took was a simple idea to spark something that defines us as a company. That something is innovation and creativity. Perhaps all the more so because we did it in a simple way – with no rollout plan!"

There should really be a new term to describe a company like CLAAS: tradinnovation would fit the bill. The world is characterised by disruptive business models where what is considered unshakeable today is challenged tomorrow. CLAAS can always – or increasingly – find an answer to this short-termism in the software sector, as demonstrated by 365FarmNet for example, or with individual innovative features such as the use of cameras to detect field edges. "But with the large machines, it's not so straightforward. Here we have to think ahead and try to imagine what needs we will have to meet ten years down the line with a combine harvester." Cathrina Claas-Mühlhäuser speaks rapidly with great concentration and is adept at expressing her thoughts succinctly. Although we have already changed the topic of conversation, in her mind she is still mulling over a problem which she feels she hasn't explained adequately. A few minutes later she returns to it. "When I say that major strategic changes lie ahead, most employees are concerned at first. We're going through this right now! We must find an even better way of describing the destination and the journey so that it is perceived as something positive."

Among the changes are ones driven by technology. "Fewer and fewer people want to work in agriculture because it is traditionally regarded as a tough job with long hours. But we are helping to make it easier and more manageable with fewer people." And of course this is already having a huge impact on CLAAS – and is expected to do so even more in future. "We will continue to earn our money from new products. There will certainly still be physical work to do on the farm in future, but innovative services are set to increase substantially."

Customers expect early warning systems, for example, which inform them when a wearing part has to be replaced. Then if possible, they'd like to be able to order these parts at the click of a button on their smartphone, or instantly book an appointment with a service engineer. Pay-per-use models will play an increasingly important role to reduce farm costs. Software

"If our only concern is who came up with the good idea first, people will quickly stop talking to one another. The only important thing is that it is implemented!"

CATHRINA CLAAS-MÜHLHÄUSER

Dr. h. c. Cathrina Claas-Mühlhäuser, chair of the Supervisory Board and deputy chair of the Shareholders' Committee of the CLAAS Group.

After training to become a commercial administrator, Cathrina Claas-Mühlhäuser studied business administration at the University of St. Gallen in Switzerland. On completing her MBA, she worked in both Switzerland and Chile for the multinational technology company ABB. Claas-Mühlhäuser has been deputy chair of the CLAAS Shareholders' Committee since 2004 and since 2010 she has also held the office of chair of the Supervisory Board of CLAAS KGaA mbH.

Her external roles include a seat on the Supervisory Board of KWS Saat AG, which she has held since December 2007. Claas-Mühlhäuser is also a member of the Advisory Council of the Commerzbank and a member of the German Committee on Eastern European Economic Relations as well as being a member of the Board of the Asia-Pacific Committee of German Business. In May 2018 she was awarded an honorary doctorate from Harper Adams University in the UK.

Cathrina Claas-Mühlhäuser is married with three chidren.

updates for new machines are already done by remote flash, which undertakes job planning and documentation with the aid of 365FarmNet. And everything has to be very carefully accounted for – because the added value of the enhanced solution also has to be paid for. "This does not remotely mean that in future CLAAS will stop manufacturing and developing machines!" Cathrina Claas-Mühlhäuser is quick to clarify. "But all products and processes are linked to one another by the flow of data and that means we must work even more closely and more effectively with one another."

So there are plenty of challenges. Would a little caution not be advisable? "CLAAS always proceeds with caution, anything else would be commercial suicide. We want to be first to the summit – but if you can't master the complexities properly, you soon run out of steam", says Cathrina Claas-Mühlhäuser. "But dithering and procrastinating will set us back. We have to switch to forward gear because the best way to meet new challenges is with new solutions!"

Behind every idea, there's a clever mind

It is innovative people in the company that make a company innovative. Right from the very start, CLAAS thrived on the ingenuity of its founders, the four brothers Bernhard, August, Franz and Theo Claas. Their spirit lives on in the company today – and it's infectious, as the following history of "Claasian" ideas shows.

GALIP SARIM

Prolific ideas generator
I just can't help myself: whenever I look at a work process, I start to consider how it could be optimised. I am motivated by changes because in most cases they lead to improvements. I have been at CLAAS since 1998 and during this time I have put forward 230 ideas for improvements. We have implemented more than half of them in the company, a quota which I am entirely happy with. **One idea even earned me the maximum bonus: 18,000 euros.** It was a suggestion as to how we could cut down on unnecessary transport materials for the Russian market. Obviously the award is a motivating factor, but I would suggest changes even if there was no money involved. Because like I said, I just can't help myself.

Galip Sarim (43) came to CLAAS in 1998 as a machine inspector. He is also company first aider, team spokesperson for training and disabled employees representative, so his commitment to the company goes way beyond his job. In his hometown of Beckum he is an official 'integration facilitator' whose role is to guide migrants through integration programmes.

JOACHIM BAUMGARTEN

Powerful, easy to operate

Combine harvesters became increasingly powerful from the mid-90s onwards. The machines achieved more – but the processes had become so complex that most drivers were no longer able to get the most from the combine harvester without assistance. At the time I was responsible for the further development of the combine harvester function group along with my colleagues and I remember the customers and distributors approaching us to ask if we could simplify the machine's control system. Now we were faced with a dilemma because we were convinced that it was this complexity that produced an unrivalled level of performance. So we came up with the idea of developing a driver assistance system which relieved the burden on the driver and automatically got the best from the machines. **The CEMOS innovation was born.** The role of my team was to develop new ideas and bundle them to create a roadmap for implementation. But how do we generate ideas? There are two important factors, firstly we have to develop a deeper understanding of all the processes taking place inside the combine harvester. In other words, everyone on the team must really know the machine inside out. Secondly, we have numerous discussions with customers, ideally on the field where the machine is to be used. Here we learn what a system needs to do to really be of help to the farmer.

Joachim Baumgarten (62) started his career at CLAAS in 1990. Beyond the factory gates, he also sees himself as a designer: he built his own low-energy house back in the 90s equipped with a photovoltaic system which enables the family to heat their home without using oil or gas.

PETRA MARTOK-MIHALIK

Smart logistics

The job of inventory management in Preproduction is to employ simple solutions to ensure that the material arrives where it is needed. Three years ago we had the idea of using a tool for this process which employees were already familiar with and that they always has to hand – namely their smartphone. In the first phase of the project we developed an SAP-based platform which enabled real-time stocktaking. In the second phase we made it possible to control initial logistics processes by smartphone. We are currently working on the introduction of the third phase in which we will use a DispoCockpit planning tool to accurately plan the capacities of employees and machines. This will enable us to guarantee precisely coordinated production processes. I am delighted that our solution has been so well received by the staff: what was initially new has now become routine. And the fact that many other CLAAS sites have shown interest in our system is a further endorsement of our innovative work.

Petra Martok-Mihalik (32) joined CLAAS in 2011 as a trainee at the Hungarian Törökszentmiklós site. She works in supply chain management and was appointed Head of Logistic Engineering Support in 2017.

PROF. DR.-ING. KARL VORMFELDE

The visionary and the engineer

Prof. Dr.-Ing. Karl Vormfelde (born 1881) was an academic, but above all a man of action. After gaining a degree and a doctorate, he worked in various industrial companies and academic fields, including several years as director of the former Bonn-Poppelsdorf Agricultural College, where he was head of the Institute of Agricultural Machinery and Physics until his death. During this time Vormfelde developed his talent for bringing innovations to fruition together with carefully chosen fellow comrades-in-arms, despite significant opposition. He worked with the Claas brothers and his assistant Walter Gustav Brenner to establish the idea of the European combine harvester: Vormfelde had seen for himself just how efficient combine harvesters had proved in the US in helping farmers with the harvest. **"We are now witnessing major radical changes to the structure of agriculture in large parts of the world",** he wrote in 1931. His goal was to design a combine harvester that could cope with the conditions in Germany and Europe. He chose to collaborate with the Claas brothers' still relatively small company on this project – since the industry leaders at the time had already turned him down. Vormfelde reckoned that CLAAS would be enlightened and willing to take risks – and he was not disappointed: the Mower-Thresher-Binder (MDB) was delivered in time for the 1936 harvest.

MARTIN OBER

Catching the wave

When I joined CLAAS in 1995, the focus was still very much on drum mowers. But in parallel with this, the first disc mowers had been developed in-house in response to the market trend, which were equipped with a third-party mower bar. We had various issues with the strength and functioning of this bar because it was not adequately tested. From then on, I couldn't stop thinking about mower bars. **Customer satisfaction is my main motivation** – and so we worked tirelessly with the supplier to improve the design. On account of its shape, however, there were still a few areas that were less than ideal. Then in 2009 we decided to develop this core component of the mower in-house and so the opportunity arose to implement a long-held vision – to form the bed of the mower bar from a wave-like single piece of steel to obtain the perfect shape. No sooner said than done, and now the product is firmly established in the mowing equipment sector under the name MAX CUT.

Martin Ober (50) is responsible for the design of various DISCO disc mowers and the mower bar at the Saulgau production plant.

JULIA CONENS AND STEPHAN NIEWÖHNER

One app for everything

There's something missing in the Claasian community: there is still no digital communication channel which actually reaches every employee. So together with colleagues from Production and Administration, we have developed an employee app. The aim is to facilitate communication between departments and across national borders and at the same time provide a practical tool for everyday use. In developing this app, we have drawn on the extensive IT expertise within the company: the best ideas come to us when we talk to colleagues in digital sectors.

Julia Conens (28) is a corporate communications consultant. Together with Stephan Niewöhner (30), Mobile Applications project manager at CLAAS since 2013, she is in charge of the employee app project.

OLIVER AUST

Prize-winning model

During the 80s and 90s, the COMMANDOR was a top-notch player on the field. Today the model 228 CS is a champion in the model-making sector. A specialist model-making magazine runs an annual competition in which its readers choose the Model of the Year in various categories. In 2018 the CLAAS miniature won the collectables/agriculture category. The winner is part of the "CLAAS Youngtimer" line which I have recently developed as product manager for Models. These Youngtimers are very popular with collectors because they bring back nostalgic memories: many of them drove these machines themselves in their younger days and their earliest memories of fieldwork are often associated with these vehicles. My job is to make as many CLAAS machines as possible that we sell on a 'large' scale also available as models or toys. I have a private collection of more than 40,000 agricultural machinery brochures of all kinds, including CLAAS brochures from the 1950s, which has helped me greatly with my work. Our models ensure that the innovative machines of the past live on, on a small scale at least.

Oliver Aust (49) completed his apprenticeship as an agricultural machinery mechanic and joined CLAAS in 2008 as product manager. He is author of the children's book "XERIONS großer Tag" (XERIONS Big Day) in which he tells the story of a young XERION through pictures.

The net of oranges

At the start of the 80s we were faced with the challenge of developing a round baler which did not have to stop to bind and eject the bales. The benefits of this sought-after innovation were obvious: it would almost double baler capacity, would be easier to operate and put less strain on the mechanics. Our first attempt to solve this problem involved fitting a pre-feeder to the baler which picked up the crop during the binding process. The problem was that this pre-feeder had to be very large because it took a long time to wrap sufficient twine around the bales. **The solution came to me on a family holiday when we were touring Spain in our VW camper van. We stopped at the side of the road to buy some fresh oranges which were packed in a long tube of netting.** On closer inspection it struck me that the net held fast on its own by clinging to the rough surface of the oranges, using the same principle as the burdock plant. Back in Harsewinkel, we picked up on this idea. After many attempts, the net wrap was born, which we have successfully introduced in many markets.

Dr Theo Freye has held numerous managerial posts during his 34 years at CLAAS. From 2007 to 2014 he was spokesperson for the Executive Board.

DR THEO FREYE

WOLFGANG JENNEN

Back out to the field again!

I started my service with CLAAS on 1 February 1961. It was a whole different world back then. But we were still innovative! I worked on the design and development of combine harvesters. What I always found particularly exciting about my work on new ideas was the field testing. I remember going to the Sudan in the late 70s to test special cutterbars – that was quite an adventure. My field testing work has also taken me to Belgium and the Netherlands, England and Denmark, Italy and Austria, Greece and the former Yugoslavia, the USA and South Africa. In short, I have travelled the world with CLAAS. During my career I have registered a total of 35 inventions. As an engineer, you feels proud when you succeed in developing a technology which did not previously exist – and which offers many benefits. But it is important not to be too naive or sceptical in your approach. For example, when we were testing the first machines with newly developed crawler track technology, my first thoughts were: no drive train will ever manage that! With the TERRA TRAC technology, we ultimately found a solution which far exceeded our expectations.

Wolfgang Jennen (81) began his career with CLAAS in 1961 after training as a metalworker. He always remembers what August Claas used to say when it was came to testing the machines: "Go onto the field and try it, I will pay for it!"

BERND HOLTMANN

Always critical, always looking for improvement

A few years ago my team was asked to develop a new, more powerful threshing unit for the new LEXION to go into production in 2020. From my experience of numerous harvesting operations, field trials and customer visits, I had lots of ideas about how to improve the existing threshing unit. For example, the pivoting concave bar, the main concave flap that can be engaged from the cab and the replaceable segment in the main concave. But it's hard work translating ideas into innovations. The conceptual development of a new threshing system requires countless field trials and tests. Many colleagues from all sectors are involved in this process, which requires a great deal of coordination. I am very satisfied with our work, but as an engineer I always remain critical and I'm always looking for further opportunities for improvement.

Bernd Holtmann (48) has worked at CLAAS since 1998. With a qualification in agriculture and a degree in mechanical engineering, he manages the functional division that deals with Threshing, Grain Separation and Cleaning.

THEO CLAAS

FRANZ CLAAS JUNIOR

AUGUST CLAAS

The Claas brothers

Remarkable personalities lie behind every great story. For CLAAS, they are the four brothers Bernhard, August, Franz and Theo Claas. When the company was founded, the oldest brother Bernhard was 28 years old and the youngest, Theo, just 16. August was 25 and Franz 23. Over 100 years ago, together they set about making their start-up one of the world's leading agricultural machinery manufacturers.

BERNHARD CLAAS

Born in 1885, Bernhard Claas was a man of few words who preferred to work in the background. He kept the company running smoothly with his exceptional powers of observation. Grievances were dealt with without fuss and conflicts resolved. Bernhard was trusted by his brothers and their employees. He is the one who impressed Professor Karl Vormfelde with the young company's achievements and thus had a positive and enduring influence on the company's development. Since he had no children, in 1935 he transferred his company shares to his brother Theo. He died on 18 February 1955 in Bielefeld.

August Claas, born in 1887, had already demonstrated his entrepreneurial skills with the founding of the company. Together with his brothers, he guided the business through ups and downs with courage and determination to create a global enterprise from a small family business. "Then we'll do it on our own", was his famous catchphrase. The visionary pragmatist and "father of the European combine harvester" was distinguished by a characteristic tenacity and belief in his own abilities. August Claas died in 1982 at the age of 95.

Franz Claas Jr, born in 1890, was a passionate engineer who created his own department in 1928: the tool shop. Here metal workers and lathe operators worked together under his direction. In all his activities with the growing company, he always insisted on state-of-the-art toolmaking equipment and one of his prime concerns was to avoid any reliance on external knowledge. On his many trips away, he was always on the lookout for technical innovations which he could use in the company. "The father of toolmaking" as he was respectfully known, was greatly admired by the workforce. He lived to the age of 74.

Theo Claas, born in 1897, succeeded in acquiring specialist knowledge at a young age and in a very short time through his involvement in aircraft construction and submarine development. At the age of 19 he was even appointed project manager for the construction of a new bridge. His commercial skills, diplomacy and ability to see the big picture even when dealing with complex processes stood the company in good stead. As sales director, he could deal with figures like no other and in discussions with his three brothers, he often put forward irrefutable, fact-based arguments. Everyone was saddened by his sudden and unexpected death at the age of 55.

NILS FREDRIKSEN

In search of the unique selling point

I feel compelled to find solutions to problems – it verges on an obsession which accompanies me in my sleep on certain projects. Patents have been filed for around 20 of the ideas that I came up with on my own or with the team, whilst others have been implemented without this step. I remember Project 207 particularly well. Helmut Claas learnt about a new continuously variable hydrostatic-mechanical transmission system on a trip to the US. He wanted to adapt it for CLAAS and launched Project 207 on 19 September 1978 specifically for this purpose. I was given the job of managing this challenging project from day one. The aim was to design a basic vehicle with tractor-like characteristics which, together with specially adapted harvesting attachments, would bring about a new, more efficient level of mechanisation. But this progressive thinking found little support among farmers. Ultimately, it failed to progress beyond the prototype stage. However, the basic machine with its special characteristics subsequently evolved into the XERION – it was well received by the market and now occupies a special position in the tractor world. This makes me very proud. At CLAAS we always wanted to find unique selling points. We definitely came up trumps with the XERION.

Nils Fredriksen (77) originally came from Mysen in southern Norway. He joined CLAAS in 1969 as a young engineer and apart from a brief interlude, was in charge of developing the XERION and subsequently the continuously variable transmission until his retirement in 2006.

NORBERT DIEKHANS

It can always be bettered

My wife says that even in my free time, I am often in 'testing mode': I like to retest things, even if they were already working in a different way – things can always be bettered. At work I always listened closely to what our customers wanted at field trials, trade shows and even in publications. This led me to mull things over in my mind and as ideas came to me, I would discuss them with my colleagues. Many of these ideas and conversations eventually found their way to Predevelopment – resulting in around 80 patents in the end. I remember one patent from mechanical engineering which was concerned with how best to set the crop speed in the discharge chute of a large harvester. And there was another involving the development of a route planning system for fieldwork which used GPS technology.

Norbert Diekhans (72) joined CLAAS in 1979. His most recent position was head of Predevelopment.

ALEKSANDR BEREZOVSKIY

Knowing what can be improved

I come up with ideas for improving workflows when I look very closely at everyday processes in the factory. If I notice something that in my view is not working as optimally as it might, I talk to my colleagues about it and asked them: **"Does it have to be like that – or could it be done differently?"** I have worked for CLAAS in the Krasnodar plant for three years now. I started my career as a paintshop technician and was promoted to team spokesperson for the Surface Treatment Centre after one year. The innovations I have developed aim to reduce paint consumption. After a few adjustments, we now get more coverage from the paint.

After just one year at the CLAAS factory in Krasnodar, Aleksandr Berezovskiy (35) was promoted from paintshop technician to spokesperson for the Surface Treatment Centre.

HEINRICH ISFORT

An idea with momentum

As a systems engineer at CLAAS, I was responsible for the designs of various models and series in the JAGUAR range from the mid-70s. We initially used a blower to transport the chopped material to the forage trailer, but the drawback of this technology was that it was expensive and consumed a large proportion of the engine output. Together with two colleagues, we set about finding an alternative and ultimately we developed the CLAAS crop flow principle which deployed an accelerator in the discharge chute to give the chopped material additional momentum en route to the transport trailer. When the patent expired, it became clear just how long-term our development was: **today virtually all our competitors' self-propelled forage harvesters have adopted the CLAAS crop flow principle.**

Heinrich Isfort (68) joined the company in 1976 as a development engineer for the JAGUAR self-propelled forage harvester. Although retired now, Isfort still works from home as a developer for CLAAS.

JOHANN AND HANS DÜCK

New tool protects fingers

We both work at CLAAS – father and son – so inevitably, after work and at weekends we find ourselves talking about work. As an assembly worker and a simulations engineer, we had noticed a problem: a great deal of force was needed to undo the dust caps on the valve blocks. you had to use a twisting action which put a strain on the fingers. **We thought: "There's got to be an easier way of doing this!"** So we developed a low-cost special tool with a lever that made this task significantly easier. With this tool the caps could be removed much more quickly and with far less strain on the fingers.

Johann Dück (60) and Hans Dück (24) moved from Kazakhstan to Versmold in the mid-90s. Johann Dück began working for CLAAS in Pre-assembly in 1996 and his son Hans joined CLAAS in 2018 as a simulations engineer in the Flow Simulation division.

ARTEMIY PARSHIN

Simple repairs for complex systems

As engineers and programmers in the CNC Machines section, we work with modern, sophisticated systems. The elements of the machines are highly specialised. Expensive instruments are normally required to repair them. Armed with a good idea and after a bit of tinkering, I managed to find a way of repairing the laser nozzles very easily using standard tools – an innovation which saves us time and money. I have also succeeded in implementing a few other ideas, for example a method which enables a robot to automatically make the adjustments for a special process inside the machine. **The idea is that the robot independently recognises the position of a tool and installs it in a bending machine. To make this work, I had to develop special grippers and containers in which the tools are stored.** I learnt innovative ways of thinking at university. There I was taught different strategies for coming up with ideas and implementing them. I use these just as naturally as a computer uses an algorithm.

Artemiy Parshin (37) is responsible for programming CNC machines at the CLAAS Krasnodar factory in Russia and develops his own tools for these systems.

WILHELM HÜGEMANN

Drives for a special market

As a quality control specialist, my job is to devise ways of ensuring that the quality of our products and the efficiency of production processes are maintained. So it's important to keep my eyes wide-open as I make my way round the vast world of CLAAS. That way, I always notice things that could be further improved. I can illustrate this by describing a short trip to Turkmenistan: in June 2018 it had been decided to install 3-wheel drive from then on in certain machines for this national market. However, after looking at the figures for repairs and wear, I discovered that the customers and sales distributors in Turkmenistan had very good experiences of the direct drive. In my view it made no sense to use a different drive system in this country. My advice to use direct drive for the new machines for this market was accepted – and then we had to move quickly: we were under significant time pressure to adapt the machines for this market.

Wilhelm Hügemann (60) trained as an agricultural machinery mechanic and joined CLAAS in 1982. He has been employed as a quality assurance specialist from the start.

REINHOLD CLAAS

GÜNTHER CLAAS

An international outlook

Alongside Helmut Claas, Günther and Reinhold Claas also contributed to the growth of the company. Together they represent the second generation of the entrepreneurial family. With their ideas, drive and commitment, they have been largely responsible for the internationalisaion of the family business since the 1950s.

Günther Claas, born in 1931, discovered early in his career during trips to the US with his father Franz Claas Junior that looking at other markets strengthened your own innovative capabilities. On their travels they learnt about the latest developments in machine tools at trade shows or visits to companies. It was very important, his father said, to keep up-to-date with the latest trends and developments. That was the only way to compete on the international market and remain at the forefront.

Günther Claas was quick to recognise the potential to expand the family business abroad. After studying in Göttingen, Hamburg and Innsbruck, he soon took on responsibilities within the company and in the 1950s worked tirelessly and with passion to set up the Spanish CLAAS distribution company. He was the first ambassador for the CLAAS brand in Spain, a task which he threw himself into whole-heartedly. But for Günther Claas, talking about an innovation was only one side of the coin. To convince customers of the advantages of a machine in the long-term, he followed his father Franz's creed, which he had picked up along the way. "Whatever you're engaged in, you have to make it credible." So son Günther worked with the CLAAS self-propelled machine under a fierce sun in Spanish fields and after several weeks of testing, confirmed that the machine had passed the endurance test even under extremely hot and dusty conditions.

With this combination of integrity in conversation and energy on the field, Günther Claas succeeded in establishing a strong sense of brand loyalty among customers and sales partners in Spain and other countries. His good command of Spanish acquired at university no doubt stood him in good stead. He maintained his close links with the Spanish market and when CLAAS Iberica was founded in 1967, he was appointed to the Supervisory Board, which he later chaired.

Reinhold Claas, born in 1931, was the third child of parents August and Paula Claas. After finishing high school, he graduated from the Technischen Universität Darmstadt in 1952 with a degree in industrial engineering. He was one of only a handful of students on this brand-new, interdisciplinary course, which combined mechanical engineering and business management.

Reinhold Claas officially joined the company in 1957 along with his brother Helmut. Balers and forage harvesters were always his main focus. In particular, he made a valuable contribution to the development and expansion of the newly constructed baler factory in the French town of Metz. In 1968, the Metz Chamber of Commerce and Industry awarded him a Medal of Merit in recognition of his services. During the course of his career, he filed patents for over 40 inventions.

Reinhold Claas was also the driving force behind the takeover of a renowned forage harvester specialist in Bad Saulgau. The acquisition of Bautz in 1969 marked a major step forward for the CLAAS Group: the company stepped up from combine harvester specialist to harvesting equipment specialist – eventually becoming world leader with the JAGUAR forage harvester. Under the guidance of long-standing managing director Reinhold Claas, the subsidiary became synonymous with successful, innovative forage harvesting technology.

Reinhold Claas also exercised his entrepreneurial skills outside the CLAAS Group: in 1958 he acquired two ball bearing companies – one with 250 employees in Fribourg, Switzerland and the other in the Swabian town of Munderkingen in Germany with 550 employees. He ran both companies until his 90s. Reinhold Claas is still a member of the Supervisory Board.

DANIEL NOIROT

Two become one

When CLAAS took over Renault Agriculture in 2003, it was not simply a matter of integrating the new subsidiary into the company to ensure economic success. The takeover also had a human component. My job at CLAAS France at the time was to combine the two different logistics systems to create a single efficient SAP-based system. This may sound like a purely technical task, but in our project team we soon realised that we would only succeed if we managed to meld the two corporate cultures. And we had to tread warily and not force change from above. **At this point I understood what sets CLAAS apart as a family business.** The prime focus of our considerations was not the short-term economic gains, but the interests of employees and customers. We managed not only to develop a new joint computer system, but also to allay the fears of the former employees of Renault Agriculture, who were soon proud to be part of CLAAS.

Daniel Noirot (68) joined CLAAS France in 1975. He started out in Marketing and Sales, then spent the next 24 years in Logistics, which he also managed. He retired in 2011.

The transatlantic voyager

As a CLAAS employee, I have crossed the Atlantic many times. I first went to the US in 2001 when the production of combine harvesters at the CLAAS Omaha (COL) production company had just begun. Over the next few years I flew back and forth to the US to support colleagues in the Omaha factory with configurating and repairing LEXION models. In 2004 I went to the US for a five-year stint, then after three years back in Germany from 2009 to 2012, I returned once more to Omaha to work as quality manager. **Through my experiences on both sides of the Atlantic, I learned at first hand just how important it is to have seamless supply processes for machine parts.** For many years supplies were organised manually, which not only took a lot of time but also led to errors and misunderstandings. So in 2015 I developed a digital shipping platform to make the ordering and supply of machines and machine parts substantially smoother. My hobby helped me greatly with this task: in my free time I do a lot of work with computers and software programming. The innovation I developed gave me the perfect opportunity to use a skill which I had developed outside work to solve a problem at work.

Jürgen Hauser joined CLAAS in 1995 and accepted his first job with the CLAAS production company in Omaha, Nevada in 2001. After that, he shuttled regularly between America and Germany, but has been based in Omaha since 2015 where he works as Quality Specialist.

JÜRGEN HAUSER

PROF. DR.-ING. WALTER GUSTAV BRENNER

The visionary and the engineer

Prof. Dr.-Ing. Walter Gustav Brenner (born 1899) was a gifted engineer and a very talented draughtsman who had the ability to put technical ideas down on paper accurately and at great speed. From 1929 to 1933 he worked as assistant to Prof. Vormfelde. At the suggestion of his boss, he then moved to CLAAS, where he earned his place in the company's history as chief engineer on the MDB. **"Like everything that grows up, combine harvester construction and development started out very small with us, very tentative",** said Brenner, reminiscing on the early days when he learned a great deal about the correlations between output, fuel consumption and threshing performance. It was this pioneering work which rapidly established an outstanding market position for CLAAS following the introduction of the first combine harvester.

SANDEEP HOODA

From South the North

In India, farmers' harvests and the CLAAS business are heavily dependent on the monsoon: if it rains enough, orders for machines come flooding in, if it doesn't, demand falls. 2013 was a particularly dry year in southern India, where CLAAS is market leader. The CROP TIGER 30 and 40 models are very popular for the rice harvest but due to the weak monsoon that year, demand for these special compact combine harvesters developed specially for the Asian market fell by more than half. Many employees were concerned about their jobs, the machines stood ready on the factory floor – what was to be done? A look at the weather data showed me that the monsoon in northern India had been heavy, the harvest conditions there were very good but our machines were not so well known in the North. **So what could be more obvious than to show potential customers in the north of the country what the CROP TIGER could do for them?** After all, the machine is very versatile and can also be used to harvest mustard seeds or in poplar plantations where cereals are grown between the rows of trees. Competitor machines can't fit between the rows because their cutter bars are too wide. The CROP TIGER, on the other hand, is ideally suited to this complex type of harvesting. With a highly motivated team and several machines, we arranged a tour of the regions of northern India. We stayed on the road for two months demonstrating the capabilities of CLAAS machines and attracting the attention of several potential new customers.

Sandeep Hooda joined CLAAS in 2008 having previously worked for automotive manufacturers. He is Head of Human Resources at the then newly constructed factory in the Chandigarh region of northern India.

The shape of progress

In the Technical Training Centre, which is located immediately between the atrium and the production buildings, the job of the industrial designers is to bring the CLAAS brand to life on every machine by referencing the brand values and to communicate the machine's innovative strength and character through its appearance. The design team apply the company's design principles to develop recognisable shapes such as the horizontal Y-shape and simulate virtual worlds which save costs and create a comprehensive overall picture.

TEXT André Bosse **PHOTOS** Lukas Kawa

The design team work to ensure that the innovative strength of every CLAAS machine is reflected in its appearance.

"The distinctive shape of the Y symbolises pure powerful intelligence. Like a clamp, it contains the power within the machine and at the same time controls it".

ALAIN BLIND

›DESIGN IDIOM
›BRAND CHARACTERISTICS
›Y-SHAPE DESIGN

On a sunny day in 2014, the Corporate Industrial Design team comprising industrial designers who are experts in interactions at the interface between human and machine were rewarded with a comprehensive overview. Director Alain Blind and his team lined up every CLAAS model in a field near the company headquarters. From the compact and manoeuvrable VOLTO to the biggest LEXION combine harvester, they assembled everything that the company's machinery fleet had to offer. "This was a wonderful spectacle which clearly illustrated the potential of a strong visual connection", reminisced Alain Blind. But the industrial designer didn't go to all this trouble just to enjoy the view. Behind the lineup was a plan. "We wanted to find out which design aspects are typical of CLAAS machines and convey the CLAAS brand values."

Not surprisingly, CLAAS Green took first place: the colour was developed in 1966 for the pilot SENATOR large combine harvester. Today it is a protected colour officially known as "Seed Green". "No other company is allowed to use this particular shade for their products", emphasises Alain Blind. Seed Green is unique to CLAAS but the design team felt that the brand colour alone did not suffice as a unifying element. "There were a few other common features", remembers Alain Blind. For example, around half of the machines had a light grey saddle element which provides a platform for the distinctive red CLAAS logo. Apart from this however, the machinery fleet presented a somewhat less coherent appearance. "It really didn't surprise us", says Blind today. "Until then the machines had largely been micro-designed; in other words, each one was individually designed without focusing clearly on the overall picture." Alan Blind didn't want it to be like this – and initiated change at macro level: "Our aim was to develop elements which could be adapted to any product", he said of the design innovation. "Shapes, colours, materials and consistency which are reflected in every machine and at the same time create an emotional connection with the brand essence."

The designers put their heads together in workshops for a few weeks and asked the relevant questions: what does CLAAS stand for? What do the machines do? What feelings should they arouse in the customer? What distinguishes us from other manufacturers? "We gradually came closer to a design idiom which reflects the essence of the company and is future-proof", remembers Alain Blind. "And when we finally discovered the horizontal Y shape, we realised that finally, here was an element that we could agree on for all models!" Like Seed Green, the Y-shape is now a registered brand characteristic. It can be seen on every new development, "and yet its character changes to suit the character of the respective product", says Alain Blind. "After all, it's important to ensure that the products do not lose their own essential traits and yet at the same time maintain a connection to the large product family." Customers want machines to have their own character. "A tractor used in vineyards should be compact and manoeuvrable. A large combine harvester, on the other hand, should combine elegance, power and endurance." The reason the horizontal Y-shape works so well as a unifying design element is clear when you look at the CLAAS flagship model: the Seed Green Y is a particularly effective design element on the latest LEXION combine harvester. "It is timeless and understood all round the world. It suggests power and dynamism and the fact that the different machine components work together together to deliver outstanding overall performance", explains Alain Blind.

As of now, the colours of the different machine elements have also been standardised right across the board: everything that rotates and represents the primary contact with the soil, like wheels or tines for example, is Lava Red – always on the move, a colour that symbolises the CLAAS brand value agility. Strong, stabilising machine parts such as the chassis and the complex machine interiors are Bedrock Grey, which suggests grounded reality. CLAAS Green embodies passion and is intended to make the machine an object of desire. "Agility, reality, passion", says Alain Blind in conclusion. "From now on every machine will visualise these three corporate design values – using a clear and transparent product idiom. And Light Grey is used exclusively as a platform for the distinctive CLAAS logo."

The design team's office is kitted out with bright, contemporary workstations to create an atmosphere that facilitates exchanges and encourage people to be open to new ideas. Large monitors are on the desks and the walls feature colour charts and machine elements. With dozens of sketches, samples and small models of CLAAS machines on display, the team can view the full product portfolio at any time. Since 2014 the industrial designers have reported directly to Group Management – a sensible move, as director Alain Blind explains: "The various divisions within the group all have their own particular interest in the machines." Sales, Production, Marketing, Group Management – they would all bring their own ideas to the table. "As industrial designers, our job is to pool these interests, evaluate them and ultimately arrive at a product which forms an authentic and tangible part of an overall portfolio." In other words, to design machines in such a way that they serve the customers well. “And that’s no easy task”, says Blind. "And

"In future virtual reality will play an important role in designing these machines. The experiential value will rise and the costs will fall".

ALAIN BLIND

The horizontal Y-shape is a signature design element used on all CLAAS machines and products.

The Seed Green Y is a particularly powerful design element on the latest LEXION combine harvester: it suggests power and dynamism.

yet it works, as long as we always keep in mind the CLAAS brand essence and take a holistic approach. Everything we do stems from this."

The traditional view that engineers give the machine a function and industrial designers then create a style that reflects this function means little to Alain Blind and his team. "Form follows function" is an old idiom from the design world. In fact, he was taught it when studying in Wuppertal in the 2000s. "We've come a long way since then", he says. "Industrial design, mechanical engineering and the customer as the user go hand-in-hand. We are in constant discussions, starting at the earliest preliminary design phase, then we turn our attention to the next product generation as soon as the machine appears in the catalogue." He sees the role of industrial designers primarily as being to reduce complexity, increase consistency, emotionalise the mechanical engineering feats of the engineers and offer the customer a well-designed, integrated package.

So how will this work be done in future? Alain Blind indicates the room opposite, at the centre of which is a shiny black box that at first sight looks like an anachronistic mock-up seat: loads of tech and cameras connected to a high-performance computer. "Only, this computer has unimaginably powerful graphics", says Blind. The purpose of the computer is to create virtual realities: when users put on the special glasses, they become immersed in a world simulated by processors. "All we need apart from the VR technology is a mock-up interior made from aluminium, wood and foam that roughly corresponds to a vehicle cab and is covered by numerous cameras that capture even the smallest movement in the room. The user climbs into the cab, puts on the glasses and has the impression of being in the simulated machine in a photorealistic environment." It's not just an amazing experience, it also saves a great deal of money by simulating the design and the human-to-machine interfaces: "Until now, we would have had to produce expensive models in order to show how a new development worked. In future virtual reality will play an important role in designing these machines. The experiential value will rise and the costs will fall. This is the future – and we're looking forward to it."

In discussion: The industrial designers translate the CLAAS brand values into a clear, transparent design idiom.

Virtual realities: Simulating design and the human-to-machine interfaces speeds up the development process.

"Industrial designers ensure that the engineers' technical developments, products and innovations in their entirety serve the customers well in terms of design and operability".

ALAIN BLIND

More precise, intelligent, digital: the future of farming

Robotics, connectivity and artificial intelligence will revolutionise agriculture in the coming decades. What changes can farmers expect? And how can everyone involved benefit? Three academics provide some answers.

TEXT BERND EBERHART **PHOTOS** Christian Perner

› BIG DATA
› ROBOTICS
› ARTIFICIAL INTELLIGENCE

The past and future of farming have been brought together in the basement of the Institute of Agricultural Engineering at the University of Hohenheim. Here staff have created a small gallery: photographs depict farmers carrying straw bales on their backs or driving their oxen across the fields; an old wooden plough hangs on the wall right next to a glossy poster advertising a Field Robot Event – twelve small four-wheel robots stand in a field. They are waiting to take part in a competition in which student teams from all round the world program the latest agricultural robots to complete a series of tasks fully autonomously. Two entirely different worlds, separated by two short steps. Is this age-old allocation of roles now set to change within a few decades? Will robots and artificial intelligence systems monitor our fields and crops in future?

Agriculture is characterised by a love of ingenuity and a willingness to embrace new technologies and processes that is virtually unequalled in any other field. And yet many farmers ask themselves these questions – and many are extremely concerned. Precision farming came on the scene in the 1990s in the form of map-based, GPS-assisted steering systems for combine harvesters and on/off section switching for sprayers. This was followed in the next decade by smart faming, which featured increasingly precise real-time sensor technology. We are now witnessing the dawn of digital farming or Farming 4.0, which will bring about sweeping changes that revolutionise the farming industry as never before. And if we set the right course, it will offer opportunities on every level – for nature and agriculture, from producers to consumers.

Hans Griepentrog is Professor of Technology in Crop Production at the University of Hohenheim. As we walk through the Institute of Agricultural Engineering's research building, he stops at an area marked out in different colours. His master students have a collection of coloured balls which are to be transferred to the appropriate colour-coded zone by "Sparrow", a 4-wheeled robot that can be equipped with laser scanners, cameras, GPS antenna and various tools. The students' job is to program Sparrow so that he can perform the task autonomously.

Even for a robot, this is a challenge. It is tremendously difficult for them to distinguish weeds from crops or healthy leaves from diseased ones – two tasks that autonomous field robots will be expected to do in future.

Machines like this are already being used in some niche sectors with high profit margins and labour-intensive, compartmentalised cropping areas such as winegrowing or greenhouse vegetable production – albeit with a very narrow, clearly defined range of tasks. The French manufacturer Naio, Swiss company Ecorobotix and British Small Robot Company are just three examples of research-based start-ups developing agro robots that are already available or on the verge of market-readiness; many similar young drivers of innovation can also be found in the area around Osnabruck known as Agrotech Valley. "This is an incredibly dynamic field", says Hans Griepentrog. "But we still have a great deal of development work to do, especially on sensors, before we can produce really robust, functioning, autonomous field robots."

For Joachim Hertzberg from the German Research Centre for Artificial Intelligence (DFKI) in Osnabruck, robotics has long been part of arable farming. "But robotics and robots are not the same thing", the computer scientist observes. In future, robotics will not be an independent, autonomous unit. Instead it will take on some of the functions of a larger machine. "We're talking about automated machines which behave rationally and purposefully under conditions that are neither controlled nor fully known", explains the director of the Plan-Based Robot Control research department at the DFKI. Drawing parallels with autonomous cars, however, is something of an oversimplification: "The car has one purpose: to drive. When it comes to agricultural machines, that's the easiest part." With such enormous potential for development, engineers in research departments everywhere are working full speed ahead.

Stefan Böttinger is also aware of the large amount of development work being undertaken in the automated farming sector. He researches agricultural automation solutions at the University of Hohenheim. "Road traffic provides us with a

"In general terms, AI simply means using clever software to solve practical problems. But one particular form of AI is incredibly effective at analysing giant datasets and recognising certain patterns: machine learning".

JOACHIM HERTZBERG, RESEARCH SCIENTIST AT THE GERMAN RESEARCH CENTRE FOR ARTIFICIAL INTELLIGENCE

structured environment", explains Böttinger. "But there are far fewer points of reference in the field. And we have to safeguard the working areas even more extensively in all directions." Although much of this is already technically feasible – Böttinger developed the industry's first GPS-guided tractor back in 1997 and built an autonomous forage harvester 15 years ago in Hohenheim – no heavy agricultural machines will be used in fields in the foreseeable future without human supervision.

Robots are without doubt among the most spectacular innovations in farming. Yet Hans Griepentrog sees robotics only as the last and highest of four stages of development in digitalised farming: "We start with machine connectivity", explains the agricultural engineer. And indeed, the Internet of Things has already found its way onto many farms: the ISOBUS communications standard digitally connects a tractor's control terminal to different implements and enables data to be exchanged with a central computer on the farm via mobile phone networks or WLAN connections. In the second stage, cloud computing stores the data securely and makes it available flexibly. "The third stage is all about analysing these data. Here we enter the realm of big data and artificial intelligence." A modern farm already produces vast quantities of data. In future, however, not only will data resolution dramatically increase, more and more new parameters will be added. "Sensor technology today is still very much focused on the crop", explains Hans Griepentrog. For instance, remote sensing systems can record the condition of the vegetation by satellite or drone, and LED-based sensors on the machines can determine the current nutrient supply of plants before fertiliser is applied. Griepentrog believes that there is still great potential here: "It would be very helpful if we could also determine the exact soil chemistry during the growing phase." At present this is difficult to do, especially in deeper soil layers.

Big data has arrived on the farm; data about soil, weather, vegetation and applications collected by satellites and sensors and from grain stores; actual values, target values, reference values, empirical values – the digital farmer seeks to extract meaningful information he can act on from this huge wealth of information. And to do this, he needs the help of AI – artificial intelligence.

"In general terms, AI simply means using clever software to solve practical problems", explains Joachim Hertzberg. "This in itself is nothing new. But one particular form of AI is incredibly effective at analysing giant datasets and recognising certain patterns: machine learning." Appropriately trained applications can draw the right conclusions from a seemingly unfathomable jumble of data. Embedded in farm management systems, they provide farmers with a concise, detailed overview of their farm and offer practical tips. Automated data collection and analysis initially offers farmers greater convenience. And agricultural machinery manufacturers can collect machine data

remotely using telemetrics, analyse them and alert the farmer to specific maintenance issues.

Naturally, farmers can also save money and protect the environment to boot by optimising production processes and tailoring nutrient and pesticide use to the needs of individual plants. But data also have an intrinsic value – which is proving highly controversial. More and more start-up companies are flooding the market with their services in order to get a slice of the big agricultural data cake, while established agricultural machinery companies offer their own farm management systems. Regarding the ownership of data, the German Agricultural Society (DLG) proposes that "All farm data belong to the farmer". After all, farmers want a share of any future profits generated by their data. "This is a very hot topic at the moment", reports Joachim Hertzberg. "The stakeholders involved are very diverse, ranging from small-scale farmers to agricultural machinery manufacturers and seed companies. Urgent clarification is needed regarding what an equitable business model might look like." In the next few years companies will be grappling not only with the development of fair participation models. Data protection and fail safety are unavoidable aspects of digitalisation as well.

The three academics are by and large optimistic about the digital future of agriculture – for everyone involved. But won't digitalisation make arable farming ridiculously complicated, with all these robots, sensors, data and artificial intelligence systems? "Nonsense", counters Hans Griepentrog. "Agriculture has always been ridiculously complicated. Nature is complicated!" We simply want to get closer to nature, explains the engineer, with increasingly sophisticated methods of analysis and artificial intelligence systems. So that people can then make better decisions.

Maximising the essentials

Frugal innovations are the current trend: lean, efficient solutions perfect tailored to customers' needs. CLAAS began applying this principle decades ago – long before there was even a word for it.

TEXT Bernd Eberhart **PHOTOS** CLAAS

CLAAS

The CROP TIGER was designed specifically for harvesting conditions in Asia.

› QUALITY
› EFFICIENCY
› CUSTOMER NEEDS

In the monsoon-flooded paddy fields of southern India, the rice growers' state-of-the-art machinery came to a standstill. Literally. Since the 1960s, India's agricultural system has been undergoing its very own green revolution. Yields per hectare increased several times over and thanks to modern seeds, effective fertilisers and motor-driven agricultural machines, farmers were able to produce food such as wheat or millet for a constantly increasing population. As one of the leading agricultural machinery manufacturers, CLAAS soon attracted the attention of this emerging agricultural nation: as early as 1969, a team of Indian engineers from the machinery manufacturer Escorts made their way to Harsewinkel to examine the latest developments in harvesting machinery.

In the 1980s, CLAAS finally began to export the DOMINATOR 80 to India. But the combine harvesters struggled to get going in the small paddy fields in the south and east of the country: the wheels got bogged down in the muddy flooded fields. Even the best, most powerful machines would still have got stuck in the thick mud between the rows of rice. A customised solution was needed if machines were to be used in the fields – and a team of developers who would take a radically different approach. Contrary to the usual way of doing things, it would not be necessary to make each new machine larger and more powerful than its predecessor. Instead, it would have to be better tailored to the special conditions. In 1993 a mini revolution made its way onto India's paddy fields: the CROP TIGER, a compact, lightweight and manoeuvrable rice combine harvester. Its unique selling point was a robust chain drive instead of wheels. A truly frugal innovation. Although nobody called it that in those days.

The Duden German dictionary still recognises the term "frugal" only in the context of "lifestyle, especially food and drink", defining it as "simple, modest". The lexicographers cite "a frugal (simple, but good) meal" as an example. We have all come across the principle described here: a clever cook who in no time at all conjures up a meal from a few fresh, tasty ingredients which leaves nothing to be desired, despite its humble origins. And of course experienced cooks know that when rustling up a quick meal, they still have to cater to the tastes and preferences of their guests.

For a long time now, engineers and designers have been doing what has been done in kitchens round the world for centuries. They are the originators of all those new ideas that don't follow the doctrine of "bigger, faster, stronger", focusing instead on the essential elements, getting rid of unnecessary and expensive accessories and offering a solution that is both intelligent and efficient. When Liza Wohlfart and her colleagues at the Fraunhofer Institute for Industrial Engineering in Stuttgart ran the first seminar on this subject in 2009, there was no established term for the concept; they simply spoke of "low-cost innovations". "One of the defining features of a frugal innovation is that there is no loss of quality", explains the academic. "Only the functional scope is reduced, and in such a way that a product is perfectly tailored to its target group." And this target group is mostly in the broad middle section of society – where there are many potential customers who nonetheless often tend to be very price-sensitive. A high-tech product with all the bells and whistles is out of the question for them on cost grounds. They prefer to invest their money in fewer functions, but these functions must be relevant and reliable – in other words a lean, frugal product.

At CLAAS the classic example in this segment is without question the CROP TIGER which today has a market share of over 50% in southern India. Having first launched the TERRA TRAC version with caterpillar tracks on the Indian market, CLAAS India subsequently manufactured a wheeled model which offered farmers the advantages of the compact, robust machine for use on different terrains. CLAAS now manufactures a similarly perfectly tailored product for the Chinese market under the brand name CHUNYU: the MC H80 combine harvester. "Many farmers in China produce two harvests a year with completely different crops", explains Shufeng Kang, head of Product Management at CLAAS in China. "In the main season they grow wheat. But in the second season they harvest a wide range of crops, from maize to millet or soy bean." So just like India, the farmers need a machine which can easily be adjusted to suit different crops.

"We adopted several basic concepts from our products for the European market to use in CROP TIGER and MC H80", reports Bernd Kleffmann, Senior Vice President of Development and Systems Engineering at CLAAS. "But then we modified and adapted them to suit the individual markets. To do this we looked very closely at the customers, the crops they grow and the size of their fields." The aim is to find out exactly what the farmers need, and in particular, what they don't need. The developers of frugal products invest an enormous amount of time and effort in these questions. "For example we send colleagues from Customer Service, Sales and Product Management and even Development to visit the farmers and ask them about our products and their needs", explains Kleffmann. "Or we invite them to special customer events to give us feedback. And we look particularly carefully at our leading customers – those customers that have new ideas or who are using a product in a new way or using their initiative to tinker with the machine and add extra functions." So over time we build up a very accurate picture of the customers' needs. Then the next generation of a machine can be tailored even more closely to their requirements and maybe designed even more leanly.

This also requires thoughtful cooperation between Development, Production and Sales. With CROP TIGER and MC H80, for example, the engineers removed as many of the sensors and electronic components as possible – not least because mechanical solutions are much easier to repair. The same applies to the solenoid valves, hydraulic components and electric motors, and even some of the purely comfort-enhancing elements were crossed off the design plans. Production also works with local manufacturers and parts. In India and China, for example, this does away with licence fees and high import tariffs, because the machines are manufactured entirely in both countries. "So we are close to our customers and their needs in every respect" says Kirpal-Singh Sian, Deputy Senior Vice President of CLAAS India.

Liza Wohlfart confirms: "Developing a frugal product is anything but frugal. It requires at least as many heads as the high-end solutions." You have to get to know your customers really well and understand what they want – and also what attracts them: "Frugal innovations often have an attractive feature which may not even be available in the top-of-the-range

Frugal product: Robustness paired with "German Technology" and an affordable price make the CROP TIGER stand out from the crowd.

Top: CLAAS manufactures machines in China under the established brand name CHUNYU.
Centre: CLAAS reisssued the JAGUAR 25 for the Indian market.
Bottom: In 2018 the DOMINATOR 370 celebrated its premiere as the first seed green combine harvester for the Chinese market.

products", explains Wohlfart. "In a budget hotel for example, that might be a really comfortable mattress. For business customers passing through, this is really important. They don't need extras like pools and saunas."

The attractive qualities of the CROP TIGER are the chain drive crawler track and its great robustness: "These vehicles are built to last for ever", says Kirpal-Singh Sian, who is in charge of the development team at CLAAS India. "Time and time again I see CROP TIGER in the fields which have been in use for well over 20 years." The model is designed to appeal specifically to customers who value quality as well as affordability. "A frugal product must be clearly distinguishable from a cheap product in terms of quality and image", explains Liza Wohlfart. A completely different example from India illustrates just how tricky it can be to get it right. Some years ago the vehicle manufacturer Tata launched the Nano model to offer an affordable car for under 100,000 rupees – equivalent to around 1500 euros. But the Tata Nano was a flop. Not just because of the quality defects, but because it had a reputation for being a cheap car. "If I go to the trouble of carefully saving my money, I want to be proud of my car", explains Wohlfart. "But the Nano didn't cut it as a status symbol. And that's because the manufacturers misjudged their customers." But due to its robustness and the label "German Technology", that's exactly what the CROP TIGER is for Indian farmers: a status symbol – not cheap, but affordable.

With an engineering degree from RWTH Aachen University, Bernd Kleffmann (right) has been working for CLAAS for 22 years and is now Senior Vice President of Development and Systems Engineering. He works with his colleague Kirpal Singh Sian to get frugal innovations off the ground – such as the CROP TIGER, which has proved very popular in the Indian market. Singh Sian joined the CLAAS team in 2019. He studied engineering at the University of Applied Sciences in Cologne.

Predict the future? Virtually impossible

Anyone who could predict events years or even decades ahead would have a clear advantage when it comes to making decisions about the future and developing innovation strategies. Unfortunately however, most respected futurologists agree that this is possible only to a limited extent; the world does not change according to plan.

TEXT IRIS RÖLL **PHOTO** JASON LEUNG

A crystal ball? No, he doesn't have one of those. Reinhold Popp smiles: "It would be nice, but I'm afraid that futurology isn't quite as simple as that." The 70-year-old ought to know, having spent half his career studying the future. He teaches at the Freie Universität Berlin and at the private Sigmund Freud University in Vienna and has written well over a dozen books on the subject. The Austrian academic winces at the term "futurologist": "We prefer to say, I'm a social scientist who is concerned with future issues."

› FUTUROLOGY
› SIMULATIONS
› SURPRISES

There are scores of self-appointed trend researchers and futures institutes, strategy advisers and eventuality analysts out there. Every week a new trend is pounced on by the media: in the future, our houses will be smart enough to control themselves, robots will have taken over half our jobs and we will be living on insects. The countryside will be bled dry and the electric pedal scooter will be the future of mobility. If you ask the general public, most agree that it's only going to get worse. In German-speaking countries, the future is regarded more as a risk than an opportunity. The "German fear of the future" is a familiar term among international academics.

That's why we have this longing for a clearer picture of the future. People want to be equipped to deal with what lies ahead. Because they have to make decisions every day – whether it be the company considering purchasing a new machine, the politician who wants to know whether or not to support a draft bill, the employee trying to decide whether a certain training course will be useful or not. What's going to happen next? What do I need to face the challenges that lie ahead with confidence? Innovation is generally the answer. And someone who can predict the future has a better idea of what these innovations should look like. So what can scientists predict? Which areas are they completely in the dark about? An insight into this profession – which is dependent on so many variables – shows that although we have a long way to go before we can achieve our objective, experts have developed astounding methods for narrowing down the uncertainty of future developments piece by piece.

So how does it work without a crystal ball? "Well, first you have to be clear about what you don't know and what you have a really good understanding of. And you have to make it clear that your predictions cannot be regarded as definitive", says Reinhold Popp. Will we still be walking around drinking coffee out of disposable cups in 15 years' time? When will air taxis be ready for series production? No one can reliably predict the exact timing of these events. "And then you need the right experts, because futurology is a highly interdisciplinary field", the professor goes on to say. In his specialist area, the future of work, these might be doctors and economists or even engineers, historians and sociologists. From these perspectives scientists then examine how people have overcome previous challenges and what changes have already been implemented. Next they feed in the results of surveys: Where do employees stand on the issue of working from home, what about companies and politicians?

Then the fine art of the futurologist really begins. For now they have to collate and evaluate these findings. What are the signs that this trend is on the rise, that one is declining while maybe this third one acts as a counterbalance? "From this process we develop prognoses and scenarios based on certain "what-if" assumptions", explains Reinhold Popp, who is keen to stress that the brave new data world will not relieve the scientists of this task. "The vast quantity of data available presents a huge opportunity, but ultimately machines cannot make predictions. This is still a job for humans, and that's a good thing because humans and chance are the essential unknowns in the equation of the future."

Technical innovations and creativity, natural disasters and important intellectual movements have shaped our society for centuries. They make the future "unimaginable" in the literal sense for present-day experts. The first major computer simulation in the field of futures studies – "The Limits to Growth" study commissioned by the Club of Rome in 1973 – was a closed system with finite resources which could only offer a series of disaster scenarios. Things have turned out differently, so far at least; the predicted scarcity of resources has triggered "induced" technical progress. More efficient extraction methods and consumer electronics have been developed. Energy conservation has become a generally accepted goal. At the same time, this positive effect is the ultimate dilemma for

"The vast quantity of data available presents a huge opportunity, but ultimately machines cannot make predictions. This is still a job for humans".

REINHOLD POPP

futures studies, and is constantly used as an excuse; both individuals and governments modify their behaviour as soon as they learn of a prognosis in order to take advantage of it. As a result, the conditions change and the prognosis is invalidated.

This is routine work for Frank Offermann. With a PhD in agrarian economics from the Johann Heinrich von Thünen Institute, Federal Research Institute for Rural Areas, Forestry and Fisheries in Braunschweig, he is continually making scientific predictions which are unlikely to ever turn out right. But that's not really the point, because Offermann and his fifteen or so colleagues' job is to provide politicians with models to help them assess the impact of their actions on the agricultural market and society; a kind of playing field where they can move individual figures around, set or remove boundaries, allocate or withdraw funds. Every two years the scientists create a baseline – a projection for the next ten years based on the assumption that policies will remain the same and measures that have already been agreed will be implemented.

The latest version to 2027 predicts, among other things, that EU agricultural exports to Asia will increase, cereal prices will be slow to recover and a significant reduction in ammonia emissions will be offset by a slight rise in CO_2 output. Agronomists also address detailed questions in special studies. Are additional state subsidies for farmers helpful? What is the long-term impact of flood damage on agricultural land? What impact do direct payments to farms have on structural change? And if the German Federal Ministry of Food and Agriculture wants to know how many million euros of subsidies are required to maintain the income of small-scale feed producers at a more or less constant level – Offermann and his colleagues can use their model to perform the calculations.

A globally connected trading world has not made their job easier. "As bilateral trade agreements become increasingly common, so too do the number of variables which one way or another must be factored in – which makes forecasting more difficult", the 49-year-old explains. And how does he deal with the anticipated technical advances? "Our projections are only for ten years. We can predict what will be relevant in practical terms for this period." He is not in the business of making long-term projections but the agronomist envisages that the major trend will be towards increasing globalisation and constantly rising cross-border trade combined with changing dietary habits and consumer demands for higher levels of animal welfare. "It will also be interesting to see how we regulate global agricultural trade when domestic consumers reject certain methods such as genetic engineering."

The type of modelling that Offermann and his colleagues from the Federal Research Institute for Rural Areas already do is something that Dirk Helbing wants to apply to the entire world; the "Living Earth Simulator" project aims to create a network of computers that will use enormous quantities of data to simulate complex processes so that we can better predict disasters and their impacts – quite simply, a computer that can simulate everything on the planet. But is he a futurologist?

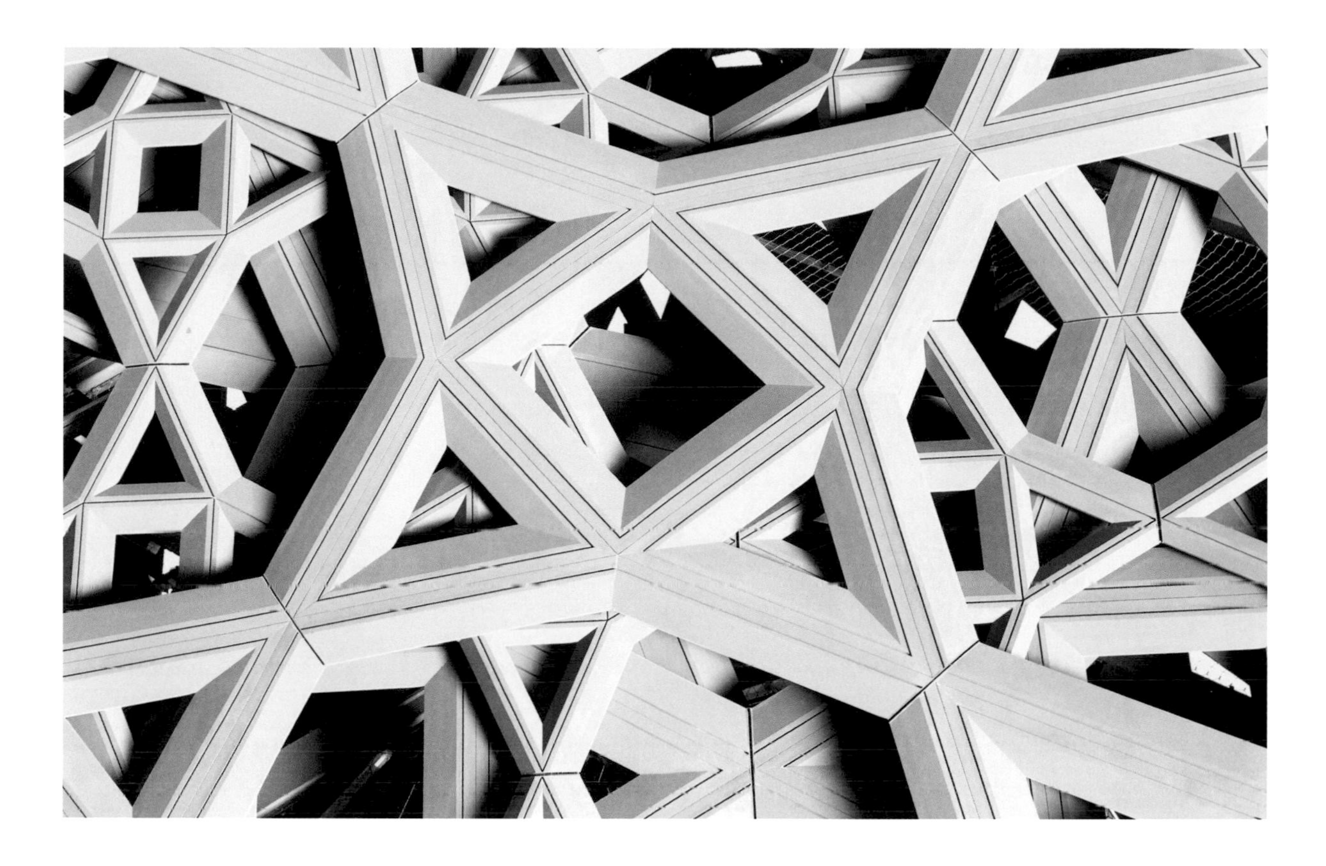

No, the professor from Zürich isn't keen on that title either. "I prefer the term complexity scientist!" The 54-year-old mathematician and physicist, who currently teaches sociology, has failed in his bid to secure EU funding worth one billion euros over ten years to implement his idea with the backing of over 100 fellow academics. "We wanted to implement the project transparently for the common good", reports the professor. "With an earth simulator of this kind, it would be possible to conduct far-reaching experiments. What happens, for example, when we change factor x, what will be the outcome of elections, what will be damaging to the competition or other societies?"

Crystal ball software which warns of cyber attacks or terrorists is already on the market. The police force in the state of Hesse, for example, is the first in Germany to deploy the Gotham programme made by US company Palantir. Within seconds it can identify links between potential terrorists which would not have been found before. It does this by analysing police databases, telecommunications surveillance data and even Facebook posts. The objective is clear – to remove dangerous people from circulation before they can launch an attack.

The US company Recorded Future is a more opaque project involving both the CIA and Google. It scans and connects ostensibly freely available Internet data and has thus been dubbed a "web intelligence engine" by some. The software claims to be able to predict not just cyber attacks, but political unrest and even the behaviour of competitors. A nightmarish concept which is far removed from the image of a traditional crystal ball-gazing fortune teller. But maybe we don't want to know exactly what lies in the future? Or at least not if it means that only some of us can benefit from this knowledge? "Predictions can never be exact because innovation and chance are disruptive elements. And that's as it should be", says Dirk Helbing. "And I actually believe that it would not be good for us to know precisely what was going to happen in the future because that would take away our freedom – as well as the pleasant surprises."

Where ideas can grow

Just a few hundred metres away from the production buildings, beyond the company grounds, is the Greenhouse: Behind the large glass facade, CLAAS employees can think in new ways, collaborate in unfamiliar configurations and try things out – sometimes together with customers, start-ups or even competitors, using open-source concepts and agile working practices. Seven members of the teams who work there provide insights into their work.

TEXT André Bosse **PHOTOS** Lukas Kawa and Marius Maasewerd

Making data useful

Data Science is all about teamwork: we work collectively to find ways of making information useful for the company.

Who are we?
The Data Science team

What do we do?
A company like CLAAS accrues an astonishing volume of data every day. Our machines communicate, our customers and dealers provide us with information, virtually every process in our production plants generates key data. However, the sheer volume of data alone does not create added value. The job of our four-strong team is to process this wealth of data and filter out the information that is of value to the company. We also try to identify new correlations, for example, by asking questions such as: what is the influence of the weather or political decisions on our business? Or: how do figures such as "total cost of ownership" change as a result of our new digital services? The days of laboriously entering figures in Excel spreadsheets are gone. Today we develop tools which deliver significantly faster and more targeted results. Colleagues from Sales or Production often come to the Greenhouse to discuss a theory with us: a marketing campaign did not produce the desired results because the relevant information did not reach the customer. Or an innovation in Production has brought about a significant increase in efficiency in that department. Our job is to use data to support – or refute – this theory. In so doing, we help the company make decisions that are no longer based on assumptions or gut feelings but substantiated by quantitative data.

What is our vision?
As a result of new platforms and machines that communicate with one another, the volume of data is set to increase to proportions that we would find unimaginable today. The challenge for us is to interpret as much of it as possible, and in the not too distant future we will do this with the help of artificial intelligence. Our job will be to make these systems intelligent in the first place.

When robots become our colleagues, our job will be to ensure that the machines' sensor systems are perfectly configured.

Robot colleagues

Who are we?
The Cobots team

What do we do?
Robot assistants are a familiar sight in many factories. The machines perform human functions – but within a clearly defined framework to prevent anybody getting in the way, since that could be dangerous. The cobot, on the other hand, works directly alongside a human. This arrangement works because the cobot is fitted with sensors and has learnt to stop a movement when it detects resistance. The great advantage of cobots is that they can perform the same tasks in Production over and over again without getting tired or losing concentration. Cobots can pick up, hold, presort and lift. We are currently testing potential applications for two cobot pairs in the Greenhouse: one goes to Production to assist as a programmable colleague, the other stays with us in our experimentation area so that we can adjust the configurations based on feedback from the production plants. We are also working with our team of four to make the cobots more intelligent, for example by fitting them with a camera so that they can accurately identify and pick up the required part from a pile of material.

What is our vision?
In the factories of the future, skilled workers and cobots can form new kinds of teams. This will not only make the human's job more agreeable, it will free them up to concentrate on higher-value activities which are more in keeping with their skills. In the near future it is conceivable that Production departments will ask us to provide a cobot that can perform specific operations. We will teach it these things, send it to the factory and then after three months we will review how the cobot has performed its job. So gradually new human-cobot teams will be created wherever they may prove useful.

The team is developing a standardised CRM system to meet the needs of customers.

Understanding the customers

Who are we?

The Future Customer Relationship Management team (CRM)

What do we do?

We are working on the introduction of a standardised customer relationship management system for the entire CLAAS Group that will replace the three parallel CRM systems currently in use. The standardised solution based on Sales Cloud software from Salesforce will help us to further improve the way we manage customer relations with the aid of targeted communication tools. The team includes not only a group based in the Greenhouse in Harsewinkel, but colleagues from other German sites as well as Poland, France, the UK, the US, Russia and Italy. This international focus is important to ensure global acceptance of the new CRM system across the entire CLAAS Group.

What is our vision?

No matter what we are doing at CLAAS – the customers and their needs are always at the centre of our endeavours. So as a company, our challenge is to get to know our customers very well. Better, in fact, than all the other players in our industry. To maintain and develop this edge in terms of customer knowledge, we are linking Sales Cloud to the Marketing Cloud and Service Cloud systems that are being introduced in the company in tandem. Our aim is to help boost efficiency in Sales in a sustainable way in order to generate profitable growth in our markets. The system will be rolled out in the different global markets in five stages, by the end of which almost 5000 employees at CLAAS and our distribution partners will be using the standardised system. It will be launched with the Polish distribution company CLAAS Polska in the first phase. Further distribution companies will follow in a staggered introduction scheduled for completion in 2022.

Learning to be playful

Who are we?
The Robotics Education team

What do we do?
Our small robotic vehicles may look like toys, but they perform an important didactic task. In a very simple way, these models demonstrate the interactions between different disciplines that also take place in our complex machines. Our robots explain the symbioses between mechanics, electronics and digital components using a simplified layout. They help the mechanical engineer to understand why software is so important today. And in turn, the IT expert grasps why even the best software comes to nothing if the mechanical and electronic components do not work optimally. We have noticed that when these symbolic robots are used in workshops, a new way of thinking very soon emerges and an interdisciplinary language develops. We run these workshops for schoolchildren from the region as well. At present, classes from the high school come to the Greenhouse three times a week and have their lessons here. It's interesting for us to observe how the young generation search for solutions. We draw on these experiences when developing innovative concepts for technical training in-house.

What is our vision?
Our machines increasingly combine expertise from many different sectors. Although our products are rooted in mechanical engineering, they would be inconceivable without software and electronics. In future it will not be enough simply to supply the core knowledge from one of these sectors. More and more, we need junior and senior staff who have a comprehensive understanding. Workshops with model robots are designed to encourage this.

Definitely not just a toy: working with robotic vehicles introduces young people to engineering.

The headset provides a gateway into virtual and augmented worlds – with a view to visualising innovations.

Discovering new worlds

Who are we?
The Virtual & Augmented Reality team

What do we do?
In Virtual & Augmented Reality we look at various application scenarios. Virtual reality immerses the user completely in a contained digital 3D world.

We use augmented reality to enrich the real environment with additional visual information, e.g. 3D models, text or symbols. With virtual reality technology we can let our developers see and experience a product even before it physically exists. Compared with a two-dimensional image on a screen, this technology gives us a more immediate understanding of size and scale and simplifies studies that focus on the person – an approach that is of particular benefit when we are examining visability and ergonomics. It simplifies the decision-making process for personnel outside the design departments. For example, customers or management can better judge the appearance or even the functions of a product even at an early stage of the development process.

While virtual reality is already used productively, augmented reality is still in the research stage. We are currently assessing the possible uses of different hardware such as headsets, smartphones and tablets.

The integration of real and virtual content offers enormous potential along the product life-cycle, but presents an even greater challenge in terms of implementation. Service departments can certainly benefit from virtual reality as a tool for servicing machines. In the same way that a navigation system can assist the driver of a car on an unfamiliar route, augmented reality can assist technicians with wide-ranging and increasingly complex repairs.

What is our vision?
We expect that virtual and augmented reality applications will increasingly complement and optimise existing working practices throughout the company. The advantages of both technologies will be so clear-cut that in future we would no more be able to imagine life without them than without our smartphones today.

"We have created a space that fosters a climate of creativity. A Greenhouse to grow new ideas".

JENNIFER LIEBIG

The relaxed atmosphere in the greenhouse provides numerous opportunities for colleagues to get quick responses and hold interdisciplinary discussions.

3D-printed components: work in the Greenhouse focuses on the printing technology, as well as innovative sales models for the components.

Shaping up for the future

Who are we?
The 3D Printing team

What do we do?
As the performance of our 3D printers increases, the speed at which we can print components – even large ones – increases. Such is the demand for these parts from the entire CLAAS Group that our printers in the Greenhouse are normally working to full capacity. We are already receiving commissions for small batches and we also experiment with new materials. Part of our job is to promote the additive 3D printing process among colleagues in Production. Most of them have learnt subtractive manufacturing, which involves removing material from a blank until the desired shape is achieved. In contrast, additive manufacturing means depositing and solidifying the material layer by layer according to a digital blueprint. This technique offers enormous freedom in terms of choice of materials and design. Parts are normally significantly lighter and it's possible to produce almost any imaginable shape. Admittedly, our parts cost substantially more to produce than mass-produced parts made by conventional means. But nonetheless, 3D printing already makes financial sense for certain applications. For example, if the tool for manufacturing an element is damaged and it would be too expensive to purchase a new one.

What is our vision?
It is conceivable that we will sell and supply scarcely any spare parts in the near future. Instead we will send them through the network in the form of a dataset. Customers will receive a file, go to their local 3D printing service provider and get the part printed out. Questions about licences and liability still need to be clarified, but this distribution of datasets has the potential to revolutionise the buying and selling of individual components.

Dialogue with partners

Who are we?
The CLAAS connect team

What do we do?

CLAAS connect is the portal through which we communicate with people who work with our machines and services. The idea behind it is to provide customers with as much information as possible about the status and performance of their machine. Once the customer has registered on the platform a CLAAS product that they have bought from a dealer, they can access data at any time. For example, exactly where the machine is, exactly how it is performing, which important fluids need topping up and when, which parts need servicing. This information is recorded by the CLAAS products themselves: since the end of 2018, all our machines have been sending data and the customer is free to decide at the time of purchasing whether this data is transmitted anonymously or whether to allow us to assign it to them. We have found that the better we explain the benefits of CLAAS connect to the customer, the sooner they are willing to share their personal data. So the development of this portal is not just a matter for IT, it also has a lot to do with customer psychology.

What is our vision?

A machine that doesn't break down – which is of course a vision of utopia. But with CLAAS connect we are working on reaching a point where potential damage does not impair machine performance at all or only very rarely. The idea is to create a maintenance system that flags up vulnerable areas even before anything goes wrong. We will offer the customer a preventive replacement service for these parts to avoid impending failure. We also plan to make CLAAS connect a tool with which customers can order and even pay for new products or services – with a view to making these processes as convenient as possible.

The strength of platforms like CLAAS connect lies in the fact that they are simple to use and function on many different levels. So the team is continually reappraising the developments.

CLAAS
8900
LEXION
CLAAS
VARIO
30

CLAAS

CLAAS
LEXION
CLAAS
CLAAS
960
JAGUAR
40

CLAAS
6900
LEXION
40
CLAAS

TABLE OF CONTENTS

CLAAS

CLAAS

G

H

I

J

K

L

M

N

O

P

Q

R

Tomislav Novoselac — **Project manager**

Marc-Stefan Andres, Edwin Baaske, Wolfram Eberhardt — **Concept**

Edwin Baaske, Lukas Kawa — **Creative direction**

Marco Brinkmann, Dr. Katrin Miele — **Project management**

Angela König, Diana Müller, Jörg Weusthoff — **Design**

Thilo Bruns, Volker Buhlmann, Ludovic Cousin, Dr. Norbert Diekhans, Nils Fredriksen, Franz Heidjann, Hendrik Henselmeyer, Heinrich Isfort, Wolfgang Jennen, Johannes Lichtenberger, Harald Lob, Tomislav Novoselac, Robert Obermeier-Hartmann, Klaus Schäfer, Christian Schulte, Werner Tertilt, Peter Weinand, Eberhard Weller, Benedikt Wiggen — **Editing**

Marc-Stefan Andres, Alexander Bank, André Bosse, Bernd Eberhart, Jörg Huthmann, Thomas Lötz, Iris Röll — **Text**

Andreas Fechner, Lukas Kawa, Marius Maasewerd, Ingo Rack, Bengt Stiller, Axel Struwe, Laura Thiesbrummel — **Photography**

Carol Finch, Leticia Pulido Alvarado, Bernard Rouvière, Fa. SeproTec — **Translation**

Arturo Castaño, Harald Katzendorn, Agnès Pokorny, Petra Schomburg, Richard Whiskard — **Proofreading**

Leonie Claas, Dr. Patrick Claas, Dr. Theo Freye, Gerd Hartwig, — **Advice**

Timo Florian Ahland, Martin Beckmann, Katrin Hameier, Michael Hartlieb, Laura-Marie Holstein, Jacqueline Osthoff, Hubert Roberg, Matthias Volk, Julia Westermeier, Andrea Wilke — **Collaboration**

Xenia Fink — **Artwork/illustrations**

Jörn Heese — **Production**

K2KONZEPT, Hamburg — **Lithography**

Kunst- und Werbedruck, Bad Oeynhausen — **Printing**

Bibliographic information from the German National Library. The German National Library has catalogued this publication in the German National Bibliography. Detailed bibliographic data can be found on the Internet at http://dnb.dnb.de.

First edition
ISBN 978-3-667-11464-8
Delius Klasing Corporate Publishing (DKCP)

CLAAS KGaA mbH
Corporate Communications
Mühlenwinkel 1, 33428 Harsewinkel, Deutschland
Tel: +49 5247 12 -3826
Email: corporate.communications@claas.com
www.claas.com

Sales
Delius Klasing Verlag
Siekerwall 21, 33602 Bielefeld, Deutschland
Tel.: 0521/559-0, Fax: 0521/559-115
Email: info@delius-klasing.de
www.delius-klasing.de

Printed in Germany

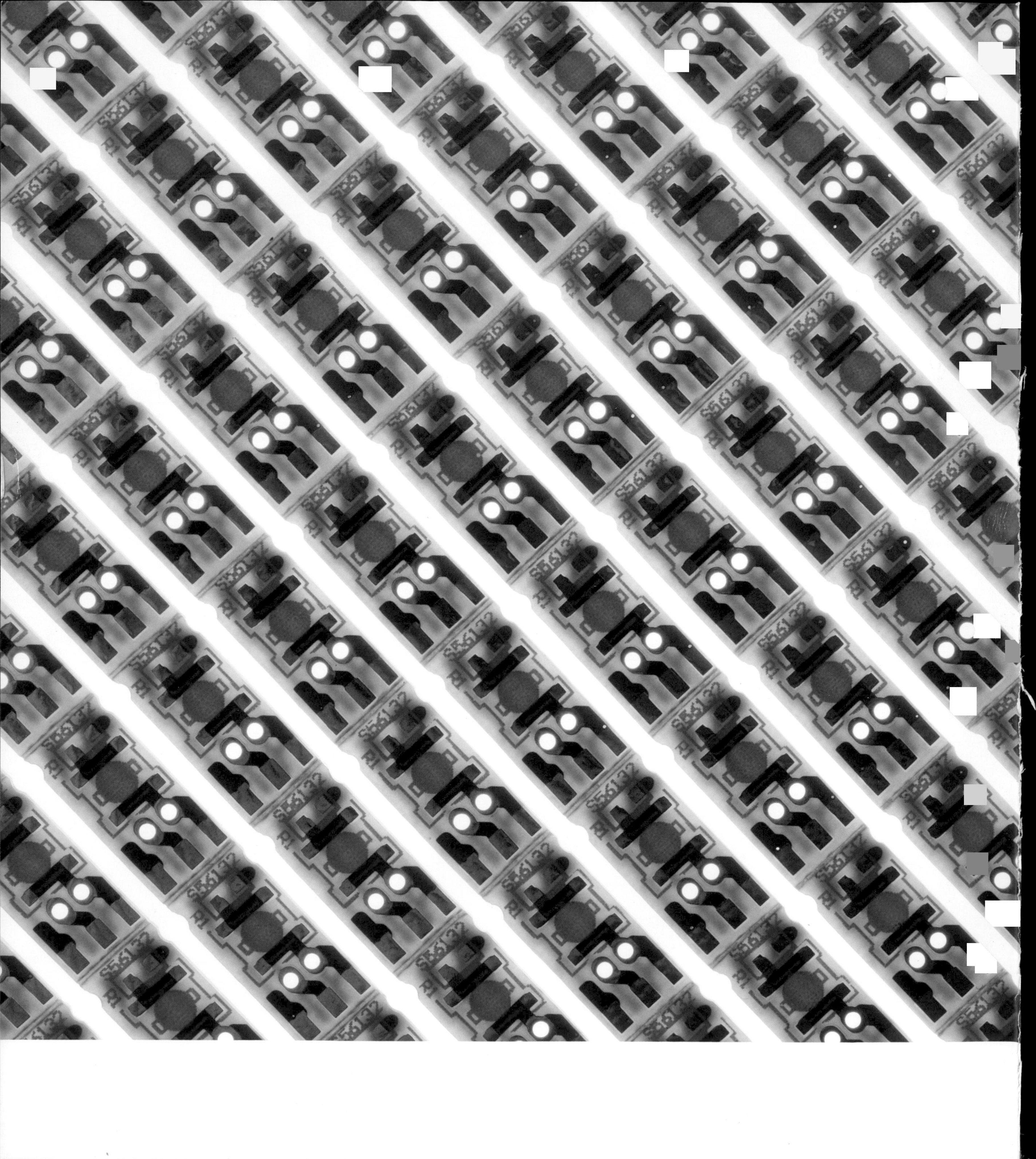